DEATH'S ACRE

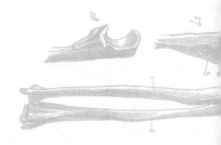

DR. BILL BASS
AND JON JEFFERSON

Foreword by PATRICIA CORNWELL

DEATH'S ACRE

Inside the Legendary Forensic Lab

The Body Farm

Where the Dead Do Tell Tales

G. P. PUTNAM'S SONS   NEW YORK

G. P. Putnam's Sons
Publishers Since 1838
a member of
Penguin Group (USA) Inc.
375 Hudson Street
New York, NY 10014

Library of Congress Cataloging-in-Publication Data

Bass, William M., date.
　　Death's acre : inside the legendary forensic lab—the Body Farm—where
the dead do tell tales / Bill Bass and Jon Jefferson.
　　　p.　cm.
　　Includes index.
　　ISBN 0-399-15134-6
　　1. Forensic anthropology—United States. 2. Forensic anthropology—
Tennessee. 3. Crime laboratories—Tennessee. 4. Body, Human—
Identification. 5. Human skeleton—Identification.　I. Jefferson, Jon.
II. Title.
　　GN69.8.B37　　2003　　　　　　　2003046908
　　614'.1—dc21

Printed in the United States of America
10　9　8　7　6　5　4　3　2　1

This book is printed on acid-free paper. ∞

Book design by Kate Nichols

DEDICATED TO

ALL VICTIMS OF MURDER,
ALL THOSE WHO MOURN THEM,

AND ALL WHO SEEK JUSTICE
ON THEIR BEHALF.

Contents

The Mayor of the Body Farm

M OST PEOPLE who attend national or international forensic science and forensic medical meetings spend more time trying to find which ballroom whatever presentation is in than the presentation itself lasts. Since I have no sense of direction, even inside hotels, I have missed my share of fifteen-minute slide shows and lectures, and arrived far too late for the handouts, too.

Missing breakfast meetings is harder to do. They are located in the dining room you eat in three times a day, and last at least an hour, usually beginning at 7:30 A.M. when everyone is tired, perhaps hungover, but enthusiastic nonetheless about seeing slides of people who were mauled to death by sharks, bears, and alligators, or killed in commercial airline crashes, or perhaps dismembered in unusual ways for unusual reasons, or committed suicide using unexpectedly creative means such as a pneumatic hammer or a crossbow. (In one sad case, when the arrow failed to kill the poor man, he pulled it out of his chest and tried again.)

The experienced and the brave ate their bacon and eggs, undaunted by the sights and sounds of gory horrors, and I was often among them, taking notes and handling myself professionally and without flinching until one rather awful early morning when the legendary Dr. Bill Bass shuffled in with boxes of slides askew under one arm and notes flapping under the other. His breakfast topic was "the Body Farm," and despite prevailing rumors that I coined that name for the only human decay research facility of its kind in the world, I didn't. The first time I met the self-effacing, funny, and brilliant Dr. Bass, I had never heard of the Body Farm. Within an hour, and without premeditation, he ruined my taste for undercooked scrambled eggs, fatty bacon, and congealing grits for the rest of my life.

"Good God," I said, appalled, early into the first slide presentation of his I had ever seen (in Baltimore, I think). "I can't believe he's showing this while we're eating!"

Dr. Marcella Fierro, the chief medical examiner of Virginia, buttered a roll and ignored me as Dr. Bass slowly clicked through one slide after another, depicting how speedily a body can skeletonize in very hot, humid weather, such as one finds in the South during the summer. I looked around the crowded room at forensic scientists and forensic pathologists, all of them buttering their rolls and stirring their coffee, some taking notes.

"My God." I pushed my plate away when Dr. Bass began to focus on maggots. "This should never be shown at a breakfast meeting!"

"Shhhhh!" Dr. Fierro nudged me with her elbow.

I avoided all such breakfasts and the Body Farm for many years. Often, scientists urged me to visit Dr. Bass's facility in Knoxville, Tennessee.

"No," I would say.

"You really should. It's not just about decomposing bodies and maggots and all that. It's about how we determine time of death, or whether a body was moved after death and where it might have been before it was moved, and who the dead person was, and how he or she died," and on and on.

Dr. Bass is jokingly referred to as the mayor of the Body Farm. In the early days of my eventual visits, just inside the razor wire–topped wooden fence, was a mailbox that the anthropologists used for leaving each other notes and messages. How odd it seemed the first time I followed that unmistakable stench of decaying human flesh and entered the acre of the dead and was greeted by a mailbox with its red flag at high alert.

"It's really not a mailbox for our residents," Dr. Bass told me rather sheepishly, as if it might enter my mind that the dead people scattered about might write home to catch up on the news. "It's just we don't have a phone out here."

They still don't. The scientists may carry cell phones, as I do, but most of us don't pull them out while wearing grubby protective gloves and perhaps rubber boots and surgical masks. When you're busy inside the Body Farm, you rarely think of calling anyone for any reason.

Throughout my career, I have emphasized that forensic experts, such as my character Dr. Kay Scarpetta, hear the dead speak. The dead have much to say that only special people with special training and special gifts have the patience to hear, despite the assault on the senses. Only special people can interpret a language very few among the living care about, much less understand.

Welcome to Dr. Bill Bass's Body Farm, the one that physically exists right this minute on a wooded patch of death-soaked land behind a hospital in the hills of Tennessee. Many of his silent guests arrive through their own selfless choosing (often making their own reservations months, even years, in advance, donating their bodies to Dr. Bass's remarkable ongoing study). Daily, wounded and worn-out bodies melt into the earth and are carried away by birds and insects and other predators who are simply part of the food chain and not the least bit morbid.

Changes to what was once human flesh can be as slight as the shift in a shadow or as dramatic as a conflagration inside one of the old, rusted cars you might find lying around at the Body Farm. Years come and go, as do the dead who have been reduced to ashes and bone, and all of Dr. Bass's patient translation adds to the fluency of a secret language that helps condemn the wicked and free those who have done no wrong.

—Patricia Cornwell

DEATH'S ACRE

CHAPTER I

The Bones of the Eaglet

A DOZEN TINY BONES, nestled in my palm: They were virtually all that remained, except for yellowed clippings, scratchy newsreel footage, and painful memories, from what was called "the trial of the century."

That label seems to get thrown around quite a lot, but in this case, maybe it was right. Seven years after the Scopes "Monkey Trial" and half a century before the O.J. Simpson debacle, America was mesmerized by a criminal investigation and murder trial that made headlines around the world. Now I was to decide whether justice had been done, or an innocent man had been wrongly executed.

The case was the kidnapping and death of a toddler named Charles Lindbergh Jr.—known far and wide as "the Lindbergh baby."

In 1927, Charles Lindbergh, a former barnstormer and

airmail pilot, had flown a small, single-engine plane, the *Spirit of St. Louis*, across the Atlantic Ocean. He did it alone, with no radio or parachute or sextant, staying awake and on course for thirty-three hours straight. By the time he reached the coast of France, news of his flight had reached Paris, and Parisians by the thousands flocked to the airfield to welcome him. The moment he touched down, 3,600 miles after leaving New York, the world changed, and so did Charles Lindbergh's life. His achievement brought him fame, fortune, and a pair of nicknames: "Lucky Lindy," which he hated, and "the Lone Eagle," which reflected both his solo flight and his solitary nature.

Five years after he flew into the limelight, Lindbergh and his wife, Anne, were living in a secluded New Jersey mansion. They had a twenty-month-old son; his parents named him Charles Jr. but journalists called him "the Eaglet." It was the heyday of sensational journalism, and savvy reporters and publishers knew that a Lindbergh story—almost *any* Lindbergh story—was a surefire way to sell newspapers. So when the heir and namesake of Charles Lindbergh was kidnapped, a media frenzy broke out: The case attracted more journalists than World War I had. The ransom notes—at first demanding $50,000, then later upping the ante to $70,000—made front-page headlines and newsreel footage; so did the claims, emerging from towns throughout America, that the Lindbergh baby had been found alive and well. But all those claims, and all those hopes, were laid to rest two months after the kidnapping, when a small child's body was found in the woods a few miles from the Lindbergh mansion. The body was badly decomposed; the left leg was missing below the knee, as were the left hand and right arm—chewed off, it appeared, by animals.

On the basis of the body's size, the clothing, and a distinctive abnormality in the one remaining foot—three toes that overlapped—the remains were quickly identified as the Lindbergh baby's. The next day they were cremated, and a brokenhearted Charles Lindbergh flew out over the Atlantic, alone once more, to scatter his son's ashes. No one called him Lucky Lindy now.

The police eventually arrested a German immigrant named Bruno Hauptmann, a carpenter whose garage rafters had apparently been used to construct a makeshift ladder used to reach the Lindberghs' second-floor nursery. Hauptmann was arrested after police traced a large portion of the ransom money to him. He was charged with kidnapping and murder: The baby's skull had been fractured, though the injury might actually have resulted from a fall, since the ladder broke during the abduction. Despite allegations that some of the evidence against him was suspect or fabricated, Hauptmann was convicted. He died in the electric chair in April of 1936.

Fifty years after the crime, in June of 1982, I was contacted by an attorney representing Bruno Hauptmann's widow, Anna. All these years after his execution, Mrs. Hauptmann was still trying to clear her husband's name. Her only chance was a dozen tiny bones. Recovered from the crime scene after the body's cremation, they had been carefully preserved ever since by the New Jersey State Police. At the request of Mrs. Hauptmann's attorney, I drove up to Trenton to see if this handful of scattered bones might somehow show that the body had been incorrectly identified—that a rush to judgment had triggered a terrible miscarriage of justice. Let them be the bones of a younger boy, an older boy, a girl of any age, she must have prayed. Anything but the bones of Charles Lindbergh Jr.

I was her final hope—a small-town scientist backing up traffic at a tollbooth as I asked directions to the headquarters of the New Jersey State Police.

It was a long and fascinating road that had brought me to Trenton, and by that I *don't* mean the New Jersey Turnpike. What had brought me here was a path that once pointed toward an uneventful career in counseling but that suddenly veered off in the direction of corpses, crime scenes, and courtrooms.

My forensic career began as a result of an early-morning traffic accident outside Frankfort, Kentucky, in the winter of 1954. On a damp, foggy morning, two trucks collided in a fiery crash on a two-lane highway. When the fire was out, three bodies, burned beyond recognition, were found in the vehicles. The identities of both drivers were easily confirmed, but the third body was a bit of a mystery.

By sheer but momentous coincidence, some months after that accident, *The Saturday Evening Post* carried an article about Dr. Wilton M. Krogman, the most famous "bone detective" of the 1940s and '50s. Krogman was a physical anthropologist who, along with two Smithsonian colleagues, virtually created the science of forensic anthropology. He was considered such a great forensic authority that during World War II, the U.S. government had him waiting in the wings to identify the remains of Adolf Hitler. As it turned out, the Russians beat the Americans to the burned-out bunker containing Hitler's bones, so Krogman never got a look at the Führer. But he had plenty of other forensic cases, from the police and the FBI, to keep him busy.

In the *Post* article, Krogman mentioned several other scientists who also specialized in identifying human skeletal remains. One of those he named was Dr. Charles E. Snow, an anthropology professor at the University of Kentucky, where I was pursuing a master's degree in counseling. The school, Dr. Snow, and I were all located in Lexington, just thirty miles from the scene of that early-morning truck collision. Although I didn't know it at the time, I was about to collide head-on with my future.

A Lexington lawyer who read the article realized that Dr. Snow might be able to identify the third victim of the fiery crash. He called Dr. Snow, who readily agreed to examine the remains. At the time, I was taking an anthropology class from Dr. Snow just for fun. When he got the lawyer's call, Snow asked if I would be interested in accompanying him on a human-identification case. This was a chance to apply, to a real-world case, scientific techniques that so far I had only read about. Why was I the one student he invited to go along? Perhaps he appreciated my budding brilliance; perhaps what he appreciated was the fact that I had a car to get us there. In any case, I jumped at the chance.

The body had been buried months before, so the lawyer completed the necessary paperwork to authorize an exhumation. On a warm spring day in April of 1955, Dr. Snow and I drove to a small cemetery beside a little country church in east-central Kentucky. By the time we arrived, the grave had been excavated and the coffin uncovered. Spring rains had raised the water table almost to ground level, so the coffin was immersed in water. As it was hoisted from the grave by a cemetery truck, water poured from every seam.

The body was burned, rotted, and waterlogged—quite a contrast to the immaculate bone specimens I had studied in the university's osteology lab. Traditional anthropological specimens are clean and dry; forensic cases tend to be wet and smelly. But they're intellectually irresistible too: scientific puzzles demanding to be solved, life-and-death secrets waiting to be unearthed.

From the smallness of the skull, the width of the pelvic opening, and the smoothness of the eyebrow ridge, even my inexperienced eye could see that these bones came from a female. Her age was a bit trickier: The wisdom teeth were fully formed, so she was an adult, but how old? The zigzag seams in the cranium, called sutures, were mostly fused together but still clearly visible; that suggested she was in her thirties or forties.

As it turned out, the police already had a pretty good idea whose body this was. Dr. Snow's job was simply to confirm or refute the tentative identification. An eastern Kentucky woman had been missing since the time of the accident; what's more, the night before the wreck, neighbors had overheard her say that she was riding to Louisville with one of the truck drivers, a man with whom she'd had a longtime relationship.

The lawyer who enlisted Dr. Snow's help had already obtained the missing woman's medical records and dental X rays. Armed with this information, Dr. Snow swiftly matched her teeth and fillings with those appearing in the X rays. By confirming her identity, Dr. Snow gave the lawyer a solid legal basis for a liability claim on behalf of the woman's surviving relatives. It seems that she and her boyfriend were killed when the other truck swerved across the highway's centerline and struck them head-on. The truck that killed them was owned by a nationwide grocery chain—The Great Atlantic & Pacific Tea Company, or A&P—so there were deep pockets to be tapped in court.

Dr. Snow's consulting fee for the case was $25; he handed over $5 of that to me for taking us to the cemetery in my car. I suspect the lawyer extracted a good deal more than that from the cash registers of A&P.

I didn't get rich that day, but I sure got hooked. It was fascinating to see the way burned and broken bones could identify a victim, solve a long-standing mystery, close a case. From that moment on, I decided, I would focus on forensics. I turned my back on counseling, switched to anthropology, and set about making up for lost time.

A year later, in 1956, I was accepted by the anthropology Ph.D. program at Harvard University. Harvard was regarded as the best anthropology department in the country, so I was honored to be accepted, but I turned them down. There was only one place to learn what I wanted to learn: in Philadelphia, at the feet of the famous bone detective Wilton Krogman.

I arrived in Philadelphia to begin my Ph.D. studies at the University of Pennsylvania in September. I was fresh from a summer job at the Smithsonian Institution, where I had analyzed and measured hundreds of Native American skeletons. I was twenty-seven years old by now—I had spent three years in the Army during the Korean War—and I had the beginnings of a family: a bright young wife, Ann—who would later earn a Ph.D. of her own in nutrition science—and our six-month-old son, Charlie. To save money, Ann and I rented a small apartment several miles west of downtown Philly.

Not long after the semester started, Dr. Krogman fell down the stairs in his house and shattered his left leg. Normally he commuted to campus by city bus, but with a hip-length cast, getting to the bus stop and clambering aboard would be nearly impossible. Since Krogman lived west of the city, too, I offered to drive him to and from work while he mended. I thought we'd be carpooling for a couple of months. As it turned out, we rode together for the next two and a half years. It didn't take him nearly that long to heal, but by the time his cast came off, I had found a new mentor, and he had acquired a new disciple.

Surprisingly, I took only one course from Krogman at Penn, but all those hours together in the car became my own personal tutorial with the world's best bone detective. It was like an automobile-age version of the Socratic dialogs, but unlike Plato, I had the great teacher all to myself.

Krogman would assign me readings, and we'd discuss them as we drove back and forth. He had a fantastic memory for authors, dates, and publication titles, as well as every detail within the articles themselves. His ability to integrate knowledge from many sources, and to apply it to solve forensic problems, was phenomenal.

Krogman didn't confine the tutorials to the car, either. Whenever he was given a forensic identification case—a set of bones from a puzzled county medical examiner or FBI agent—Krogman would call me into his lab. He would examine the bones first and formulate his analysis, but he would say absolutely nothing. Then he would ask me to look at the bones and draw my own conclusions. Then, as we compared findings, he demanded that I support and document my statements by citing recent scientific articles on the subject. Krogman was always surprised when I found something he'd overlooked. It didn't happen often, but when it did, I glowed with pride.

Krogman's teaching method was remarkably effective. Not only did it help me retain the material, it also prepared me to face courtroom questioning by hostile lawyers—something I've had to do many times in the subsequent years, though I couldn't have foreseen it then. At the time, all I knew was that Krogman was guiding me, case by case and bone by bone, down a marvelous path.

All too soon the path forked. I left Penn to take a nine-month teaching post at the University of Nebraska in January of 1960, followed by eleven years at the University of Kansas in Lawrence. But my association with Krogman was far from over. We always stayed in close touch, personally and professionally. And when I trotted up the steps of the red brick headquarters of the New Jersey State Police in June of 1982, I found myself walking in Wilton Krogman's footsteps once again.

Krogman had been asked by the New Jersey attorney general to examine the bones five years earlier, in 1977. Because of the lingering questions surrounding the Lindbergh case, the state was considering reopening the investigation. On the basis of Krogman's findings, they chose not to. Now I was revisiting that same issue on behalf of the convicted killer's widow.

By now I had attained a measure of professional standing of my own: I was the head of a thriving university anthropology department at the University of Tennessee in Knoxville, as well as the creator of what would come to be called "the Body Farm," the world's only forensic facility devoted to research on human decomposition. I had been named a fellow of the American Academy of Forensic Sciences and was serving as president of the organization's physical anthropology section. I had examined thousands of skeletons and assisted with more than a hundred forensic cases. And yet, despite all that, I felt nervous and small: a pygmy walking in the footsteps of a giant. I would be only the second anthropologist ever given permission to examine the famous Lindbergh bones.

I was ushered into a basement room of the state police building. A few minutes later a clerk brought me a cardboard evidence box. Inside were five glass vials. One of the vials had cracked at some point; it was held together with clear tape. Originally these vials had kept expensive cigars from going stale. Now, sealed with cork stoppers, they guarded a dozen tiny bones against loss or breakage—bones that represented both the premature death of innocence and the final hope of an aging widow.

Two of the bones were clearly animal in origin: a two-inch piece of rib from a good-size bird, perhaps a grouse or quail, and a small vertebral arch, probably from the same bird. Both of these bore tooth marks on them—possibly from the same dog or dogs who had gnawed off the hands of the dead child hidden in the forest.

Of the ten human bones, the largest of them—the calcaneus, or heel, of the left foot—was about an inch and a quarter in diameter; to the untrained eye, it could have passed for a piece of gravel. Four of the bones were from the left foot; two were from the left hand; and four were from the right hand. Despite the passage of half a century, decayed tissue, dirt, and even a few hairs still clung to several.

Intact and undamaged, the bones bore no signs of trauma, no indication of cause of death. The only skeletal evidence that had pointed to that—the small fractured skull—had been cremated within hours after Charles Lindbergh identified the body as his son's. What I held in my hands—these ten small bits of hands and a foot—had been sifted from ten baskets of leaves and twigs raked up from the forest floor in the days after the body's discovery. The police had hoped to find answers—a murder weapon, a set of fingerprints, something that might point to who had stolen the child and what had gone wrong—but this handful of small bones shed precious little light.

Fifty years later they still illuminated little. In childhood, skeletons are androgynous: There's no way to determine the sex of a skeleton; all you can do is measure and compare the bones you're examining with the size and development of other, known specimens. To that end, I'd brought along the two definitive reference books on this subject, *Radiographic Atlas of Skeletal Development of the Foot and Ankle* and a companion volume, *Radiographic Atlas of Skeletal Development of the Hand and Wrist*. Both represented careful studies based on X rays of hundreds of children's hands and feet. According to the measurements in those studies, the hand and foot bones from the glass vials were slightly larger than those of an eighteen-month-old male and slightly smaller than those of a twenty-four-month-old male. It took less than an hour for me to reach the same conclusion that my mentor, Dr. Krogman, had reached five years before me: There was nothing in the bones themselves to refute the notion that these were all that remained of a Caucasoid male child, aged twenty months. A twenty-month-old Caucasoid male child named Charles Lindbergh Jr.: the Eaglet.

As I slipped the bones back into their glass vials and pressed the cork stoppers tight, I was struck by how little was left—how little to mark the loss of that glittering promise, the bright future, that Charles Lindbergh Jr. could have had; the relationship he might have forged with his famous father; the pride the elder might have felt as his son grew and perhaps spread his own wings, piloting airplanes or jets or even spacecraft.

By 1982, I had three healthy sons of my own, ages twenty-six, twenty, and eighteen. I could scarcely imagine what it must have cost Charles Lindbergh in his soul to lose a young son to a violent death. But I did know what it cost to lose a different loved one to a violent, untimely, and senseless death, and I knew how fast such a thing could happen: A makeshift ladder in New Jersey breaks, and suddenly a kidnapping becomes a murder. Or a bright young lawyer's index finger curls around a trigger, and a bullet leaves a swift smear of carnage across a different set of lives. Across *my* life.

It happened in March of 1932—by utter but strange coincidence, the very same month Bruno Hauptmann was nailing together a crude ladder that bore a fatal flaw. I was three and a half years old, twice the age of the Lindbergh baby. My father, Marvin, was an up-and-coming young attorney in the town of Staunton, Virginia. He was bright and good-looking; he was married to his childhood sweetheart, Jennie (twenty years earlier, they'd been crowned King and Queen of the Maypole); and he looked to have a promising future in politics. He'd already made one run for the office of commonwealth's attorney; he didn't win, but at age thirty he still had plenty of chances—or so everyone thought.

We lived in a two-story white house on Lee Street, a couple of miles from the center of town, beside an apple orchard. My recollections of that time are few and fuzzy, but one memory of my father—of my father and *me*—remains crystal-clear: It was a Sunday morning, and he and I drove into town in our big black Dodge to buy a newspaper. (He'd come of age during the heyday of the Model T, but he'd also heard his father say, countless times, that Fords were made of tin, "and damned sorry tin at that.")

The Dodge stopped at a street corner where a man stood beside a stack of papers. Daddy reached across me, rolled down the window, then handed me a dime and asked if I would pay the man. For some reason—fear? shyness?—I shook my head no and pressed my body against my father's. He smiled good-naturedly, took the coin back, and gave it to the vendor.

I have photographs of this handsome young lawyer I'm named after. In some of them he holds me on his lap. In others he stands beside my mother. In most of them he is smiling. We were happy—*he* was happy—in those days, to the best of my memory.

But the best of my memory isn't nearly good enough, because it doesn't begin to account for what came next. One Wednesday afternoon, not long after our Sunday newspaper excursion, my father closed the door of his law office and shot himself. It was early spring; the apple trees in the orchard would have been about to bloom; U.S. farm prices were finally on the rise; and my father put a bullet through his head.

Decades later, in the one brief conversation we ever had about my father's suicide, my mother intimated that he'd been asked to invest money for some of his law clients and had lost it when the stock market crashed. Perhaps he was unable to face the people whose money he'd lost, or perhaps he was unable to face himself; who can say? Looking back on it today, when I am forty years older than he was when he killed himself, I can't help thinking, *You could have gotten past it. If you'd just hung on a little longer, things would have worked out all right in the end.* But for whatever reason, he couldn't see or feel a way, *any* way, to hang on. And so he let go.

The instant he pulled the trigger, my father slipped from my grasp—slipped away from all of us—and he remains out of reach to this day. I still miss him. I imagine the things he and I would have done together as I grew up. I long for fatherly and lawyerly advice when I'm heading into a murder trial to face hostile questioning on the witness stand. I'm in my seventies, but I still cry like a child when I recall how I shrank from paying that corner newspaper vendor. If only I had paid the man! Perhaps that would have pleased my father; perhaps he would have smiled at the bravery of his little man, felt his heart lighten a bit, felt his own courage ratchet upward one small, crucial notch.

Ironic, isn't it? Touched by death at such a tender age, you'd think I'd have had my fill of it early on and spent the rest of my life carefully steering clear. And yet, I deal daily with death. I have spent decades actively seeking it out; I immerse myself in it.

Perhaps I'm trying to prove my bravery even now, across the gulf of years and mortality that separates us. Or perhaps when I grasp the bones of the dead, I'm somehow trying to grasp *him,* the one dead man who remains forever elusive.

Sitting in the basement of the New Jersey State Police headquarters on that day back in 1982, I found nothing in those five cigar vials, nothing in those ten small bones, that could tell me anything about the Lindbergh baby I hadn't already known. Nothing to refute the evidence presented at Bruno Hauptmann's murder trial. Nothing to vindicate that half-century of hope in the heart of his widow.

Anna Hauptmann, too—like the Lindberghs, and like me—had lost someone dear. Cherished husband but convicted killer, he would continue to elude her until that day when she herself slipped from those around her, finally catching up to the man she'd lived with and loved.

Perhaps on that day she finally, fully grasped him. Perhaps one day soon I'll elude those who live with and love and should know me, and in that moment I'll find my long-lost father.

In the meantime, I search for others among the dead. From ancient Indians to modern murder victims, I do reach others. Thousands and thousands of others.

CHAPTER 2

Dead Indians and Dam Engineers

THE SKY ABOVE the South Dakota plains was a deep blue, darkening almost to purple at the top. To the west, towering cumulus clouds dropped ragged gray curtains of rain, which evaporated long before reaching the ground. From two miles above the ground, I could scan a huge expanse of rolling prairie out the airplane window. The grass and brush were already mostly brown; the Missouri River was even browner, meandering muddily into the landscape from the northwest and meandering out, even more muddily, to the southeast. The only patches of green, I had heard, were small circles of lush grass dotting the hills along the riverbank somewhere to the north of us, marking the site of an ancient Arikara village. It was the summer of 1957, a vast new horizon was opening before me, and my excitement was building.

Then, as the engines throttled back and the Frontier

Airlines DC-3 began lurching downward through the turbulence, a new sensation began building: motion sickness, my lifelong Achilles' heel. Mercifully, my flight touched down before my breakfast came up.

We landed in Pierre late in the morning. The handful of passengers ducked through the oval doorway in the fuselage, clambered down the stairs, and headed into the whitewashed one-room terminal. I looked around for Bob Stephenson, the Smithsonian archaeologist who had promised to pick me up. He was nowhere to be seen. Soon the other passengers were gone, and I found myself in an empty waiting room far from home.

The airport control tower resembled a tree house on stilts. After a while I climbed up to ask the controller if he knew the archaeologists who were working outside town, explaining that Dr. Stephenson had promised to pick me up and take me out to the site. "Oh, he's probably stuck in the mud somewhere," the controller said. "We had a lot of rain last night, and things get pretty slick around here when it's wet." Late that afternoon Bob showed up, apologetic and covered with mud. Sure enough, he'd been stuck for three hours. Little did I know it at the time, but I was about to get stuck here—of my own free will—for the next fourteen summers.

I had been brought to South Dakota by the combined might of the U.S. Army Corps of Engineers, the Smithsonian Institution, and the earth's last ice age (which ended, I might add, somewhat before my time). Twenty thousand years ago a thick sheet of glacial ice swept relentlessly southward across America's Great Plains. Shoving mountains of earth and rock before it, grinding stone into powdery alluvial soil, it reshaped millions of square miles of the planet's surface.

Now an equally relentless army of engineers, archaeologists, and anthropologists had descended on the prairie to make a few changes of their own. The engineers were starting to flood it; the rest of us were frantically excavating it, digging and sifting for buried treasure—archaeological treasure—in a desperate race against the rising waters of the newly dammed Missouri River.

The Missouri may be the most underrated river in the world. Here in America, it plays second fiddle to the Mississippi, and that's a gross injustice, in my opinion. Don't get me wrong: The Mississippi is a great river. Flowing 2,350 miles from Minnesota's Lake Itasca to the Louisiana delta, the Mississippi is a mighty waterway coursing through the very heart of America.

It's the *name* of the thing that seems unjust. Consider a drop of Minnesota rainwater that plops into the Mississippi's headwaters at Lake Itasca: From the lake's rocky outlet—small enough to wade across—that drop flows 2,350 miles before it enters the salty shallows of the Gulf of Mexico. By contrast, a Montana raindrop, falling into a spring on the eastern slope of the Rocky Mountains, journeys 2,300 miles in the Missouri River just to reach the great confluence with the Mississippi at St. Louis; from there, it continues another 1,400 miles before it reaches the Gulf—a total distance of 3,740 miles. Only the Nile and the Amazon flow farther. So, on the basis of length, at least, the Missouri should be considered the main river and the Mississippi the tributary.

The Missouri is amazing in another respect as well. To the best of my knowledge, it's the largest river that has ever changed its mind, or its destination, on a continental scale. Before the last ice age the Missouri actually flowed northeast into Canada, emptying into the icy waters of Hudson Bay. Then, when the glaciers swept down like mighty earthmovers to reshape the land, the Missouri saw an opening and veered southward, running for the warm waters off Mexico and ending up some 2,000 miles from its original outlet.

Over the ages, the Missouri has witnessed dramatic changes in the life-forms inhabiting its vast watershed. A hundred million years or so ago, dinosaurs ranged across Montana and the Dakotas. They were succeeded by a host of warm-blooded creatures, including cheetahs, camels, woolly mammoths, and huge saber-toothed cats. We humans are relative newcomers: The first inhabitants of the Great Plains might have crossed a land bridge from Asia some 12,000 years ago.

For millennia these aboriginal Americans led a nomadic existence.

Then, about 2,000 years ago, most of them began raising food crops and putting down roots. They built villages of earth lodges: round structures dug into the ground, topped with a domelike wood framework, then covered with earth and sod to insulate against the prairie's scorching summers and frigid winters. Today we'd call that "earth-sheltered housing." The Plains Indians just called it "home."

But the earth-lodge villages weren't sustainable. Trees are scarce on the prairie. They grow mainly in the river's lowest floodplain—what's called the "first terrace"—so after a generation or so the riverbank for miles upstream and downstream of a village would be stripped bare. The women, whose job it was to gather fuel and building materials, had to walk increasing, exhausting distances for wood. Eventually they would put their weary feet down, and the tribe would resettle a few dozen miles upriver or downriver in a fresh stretch of cottonwoods. A hundred years later, once the floodplain had reforested, they might circle back to the site of a village their ancestors had abandoned.

By the 1700s, the Great Plains were home to numerous Indian tribes. Four major tribes inhabited and fought over the northern Plains: the fearsome Sioux, who remained nomadic, and the sedentary Mandans, Hidatsa, and Arikara. In what is now central South Dakota, the Arikara built immense earth-lodge villages encompassing hundreds of family houses and large ceremonial lodges.

Then came the wave of the future: white explorers and fur traders. Lewis and Clark were among them, though they were far from the first. When the Corps of Discovery dropped anchor at a Mandan village in 1804, they were met by blond-haired, blue-eyed Mandans—the offspring of native women and French explorers or trappers.

On their journey upriver into the newly acquired Louisiana Territory, Lewis and Clark attempted to unite the Arikara and the Mandans in a three-way alliance with the U.S. government to oppose the Sioux, but the Arikara resisted the coalition-building and in fact skirmished briefly with the expedition as it continued upstream. The explorers fared much better with the Mandans: The Corps of Discovery wintered over with the Mandans that year, trading and hunting with the Mandan men and sharing the sexual favors of the Mandan women. Often this was done with the encouragement of the women's husbands, who believed that their wives would receive, and then transmit, the whites' "magic." Unfortunately, what was usually transmitted was syphilis.

On their return downriver in 1806, the Lewis and Clark expedition again clashed with the Arikara; in 1809, Meriwether Lewis—during an ill-fated term as governor of the Louisiana Territory—sent an army of some five hundred whites and Indians back up the Missouri with orders to exterminate the Arikara if they were spoiling to fight.

But for all their bravado, the Arikara were teetering on the brink of extinction. Within half a century of Lewis and Clark's expedition, the Arikara had all but vanished: victims of the Sioux, the settlers, and smallpox. The tribe's decimation left behind, on the second and third terraces of the Missouri, hundreds of empty earth lodges and thousands of occupied graves.

In 1957, as the last traces of the Arikara civilization were about to slip beneath the waters of progress, the Smithsonian Institution sent me out to help excavate as much as possible in the little time remaining.

THE NATIONAL MUSEUM of Natural History is one of the great Smithsonian museums lining the Mall in Washington, D.C. On the main floor, beneath its huge rotunda, an enormous African elephant stands sentinel. Several floors above him—on balconies ringing the rotunda's fourth, fifth, and sixth floors—cabinets and drawers and shelves brim with Native American skeletons. Or at least they used to.

Today, our thinking about excavating graves and collecting bones has changed radically. In 1990, after intense lobbying by Native American tribes, Congress passed a law that forbids the collection of Native American skeletal remains. The law also requires that museums and other institutions return Native American remains if those remains came from a tribe that still survives. The underlying philosophy is simple: The remains of the dead are sacred relics, not collectibles or exhibits, and they should be returned to their ancestral lands and buried with reverence. Spiritually, it makes perfect sense.

Scientifically, though, excavations and collections such as the Smithsonian's have played a crucial role in illuminating the history, culture, and evolution of humans in general and Native Americans in particular. By comparing bones from thousands of individuals, scientists can draw an accurate picture of North America's native inhabitants: their size, their strength, their diet, their average life span, infant mortality rates, and a wealth of other information. And in the latter 1950s and early '60s, those bones were pouring into the Smithsonian faster than the museum's scientists could process them.

That was lucky for me.

I HAD DISCOVERED anthropology during my last two years of undergraduate school at the University of Virginia. By then I had completed most of the requirements for my major, psychology, and finally had a few slots open for electives. As I scanned the course offerings, the first thing that caught my eye was "Anthropology." (Not surprisingly, the list was alphabetical. If I'd started reading at the bottom instead of the top, I might have wound up as a zoologist!)

Virginia didn't actually have an anthropology department—just one lone professor, Clifford Evans, who was lumped into the sociology department. But Evans was an adventurous field researcher and an inspiring teacher. He had recently returned from excavating a prehistoric village in Brazil, and his slides and stories brought its ancient inhabitants back to life in the classroom. I took every class Evans taught.

In the spring of 1956, as I was finishing my master's degree in anthropology at the University of Kentucky, I wrote Evans to tell him. I figured I was probably his only student ever to earn a graduate degree in anthropology, and I thought he might be pleased to know. By then he'd left Virginia and taken a job as a curator of archaeology at the Smithsonian.

Evans wrote back immediately. He remembered me well and told me he was glad to hear of my progress. He also told me the Smithsonian was desperate for help analyzing the flood of Native American skeletal material that was pouring in from the Great Plains, and offered to get me the job. It was a golden opportunity at a remarkable time.

The flood of bones had been unleashed by the U.S. Army Corps of Engineers. The Corps had been created to wage war on flood-prone rivers, and it did so with a vengeance. By the late 1940s its engineers had dammed and diked most of the Mississippi, so they branched out to other rivers. In the 1950s they were working their way up the Missouri.

By the time they reached the center of South Dakota, they were working on a colossal scale. Six miles upstream of Pierre (pronounced "pee-AIR" by the French but "peer" by South Dakotans), they began piling up a ridge of earth nearly 250 feet high and almost two miles long. The Oahe Dam, named for a Sioux council lodge, was the largest earth-fill dam in the United States when it was begun in 1948. It still is.

The reservoir it would create was also going to be enormous. Destined to stretch upriver about 225 miles and spread some 20 miles at its widest point, Lake Oahe would be one of the largest artificial lakes in the United States. It would inundate hundreds of square miles of prairie—and countless Native American archaeological sites.

The Corps of Engineers had earmarked part of the dam's construction cost for archaeological research and excavation, and contracted with the Smithsonian to do the scientific work. The funding was a tiny share of the dam's budget—just one-half of one percent—but the dam and its budget were so big that, by typical archaeological standards, the Smithsonian River Basin Surveys (as the overall project was called) was grand of scale and deep of pocket. As the Corps of Engineers began piling up earth to hold back the river, a small army of archaeologists and their indentured servants—undergraduates and grad students—began excavating in the area to be flooded. They began at a major Arikara site just upstream of the dam, since it would be the first to be submerged. It was called the Sully site, simply because that was the name of the

county where it was located. On the second terrace of the Missouri—
the shelf lying just above the river's floodplain—the Arikara had built
the largest earth-lodge village that has ever been discovered.

The main clue to the site's archaeological richness was a series of circles, ranging in diameter from eighteen to twenty feet all the way up to
sixty feet. These marked the locations of earth lodges; when the lodges
burned or collapsed, they left shallow depressions in the prairie, because
they had been dug several feet below grade. Rainfall is scant in this area,
averaging just fifteen inches a year, so the depressions, which collect
runoff and groundwater seepage, became tiny oases of green in the
brown prairie. (Another five inches of annual rainfall, and the plains
would have become forest instead of grassland.) The smaller green circles represented hundreds of houses, each occupied by as many as fifteen to twenty people; the handful of large ones marked community or
ceremonial lodges.

Like many of the Arikara earth-lodge villages, the Sully site had been
occupied multiple times, beginning around A.D. 1600. It was abandoned
once the nearby trees had all been cut, then resettled after the riverbank
had reforested. By dating the artifacts they found, the archaeologists
would deduce that the village had been inhabited at least three times
before being abandoned permanently around 1750.

From the ground, the earth-lodge depressions were harder to see but
easy to feel: Driving across the prairie in a jeep or truck, a farmer or an
archaeologist might feel the vehicle drop down into the slight depression, then climb back out again. The Sully site contained so many of
these depressions, driving across it was like one big roller-coaster ride.

Because the village was so big, and had been occupied for so long,
the archaeologists were unearthing a treasure trove of materials: cooking
utensils, farming tools, weapons, jewelry, and bones—thousands upon
thousands of bones, far more than the Smithsonian's handful of physical anthropologists back in Washington could sort and measure.

That's where I had first entered the picture, walking past that
stuffed elephant beneath the rotunda and up into my first summer of

bone-cataloging. A lowly graduate student, with no telephone, no pet projects of my own, no journal articles to write or review, and none of the other distractions confronting a loftier scientist, I could analyze bones from dawn till dusk. And so I did, for all of one summer and most of the next. Late in the summer of 1957, the project's director summoned me to South Dakota.

I had never been west of the Mississippi before, and I had never even flown before, so the trip to South Dakota opened up a vast new world for me. Some lessons awaited me in old bones hidden in the earth. Others were imparted by the young students who toiled in the heat and the dust of the Missouri River terraces. Still others were taught by the ants and the rattlesnakes that burrowed into the plains with us. Every one of these lessons would serve me well in the years ahead as I began applying the secrets I learned from the long-dead to understanding the stories of the recently murdered.

BY THE TIME I arrived in South Dakota in August of 1957, the summer was almost over. In just two weeks the project would shut down so the professors and students could return to school. And in those two short weeks Stephenson hoped I could help answer a question that had been puzzling and frustrating him for the past two years: Where had the Arikara hidden their dead?

From the number of earth lodges being excavated, he knew the population of the village had numbered in the hundreds and that it had been occupied for decades. But so far Stephenson's crew had managed to find only a few dozen sets of remains. So where were the rest?

Some Indian tribes, including the Sioux, put the bodies of the dead on elevated scaffolds to decompose in the open. It's therefore rare to find an old Sioux skeleton, because the bones are often scattered by coyotes, vultures, and other scavengers. The Arikara, though, seemed consistent in their burial practices. The graves were usually dug by the women, digging with hoes made from the scapulae, or shoulder blades, of bison. It was tough work with a primitive tool; so, to keep the task manageable, they made the graves as small and compact as possible: They dug a round pit about three feet deep—smaller if the individual was a child or woman—and lowered the body into a flexed or fetal position, with the knees drawn up to the chest and the arms crossed. Then they filled in the pit; covered the top with sticks, logs, or brush to deter scavengers; and topped the wood with soil and sod.

By August of that second summer, Stephenson's frustration was intense. Not only were the remains they'd found insufficient to account for the village's population, they were also insufficient to teach us much about the Arikaras' life and death. Stephenson was smart enough to know there must be an Arikara cemetery somewhere nearby. But if we didn't find it soon, we'd lose our chance.

Archaeological digs are based on a grid pattern: A site is marked off into five-foot squares, which are excavated by removing very shallow layers of soil one at a time. Each grid is assigned an identifying number, so that as the dig progresses from one square to the next, the artifacts or remains found can be logged precisely according to which square they were found in and where within the square, both horizontally and by depth. It's orderly, it's precise, and it's maddeningly slow—sometimes taking a week or more per square—so that an entire summer can be spent excavating an area just forty to fifty feet square. We had to cover lots more ground in lots less time. Stephenson put me in charge of a crew of ten students and urged me to find the Arikara dead before the end of the month.

It's hot as blazes in South Dakota in August, and the prairie is a mighty big place to search. To do the job swiftly, we'd need a small army of workers. What we had, it turns out, was a very large army of very small workers: the ants burrowing into the prairie by the billions.

The soil of the Great Plains is called loess. Pronounced "lurss," it's from a German word meaning "loose." Fine as flour, it's what put the dust in the Dust Bowl. That's in its dry state, of course; just add water, and its character changes drastically. Wet loess is quite possibly the slickest substance in the universe, and if it's sitting atop wet shale—possibly the second slickest material on earth—things get really interesting: In utter defiance of the laws of physics, friction (and therefore traction) can vanish entirely. That's why poor Bob Stephenson was so late when picking me up that first day.

Loess is tailor-made for ants. It's soft and easy to dig through, but it holds together well, so once a worker ant has tunneled through it, he can be pretty sure his tunnel is not going to collapse anytime soon.

Even better than virgin loess, in the opinion of our industrious ant, is loess that's been disturbed and loosened—for example, in the process of digging and filling in a grave. *This is nice, easy digging down here*, he thinks when he burrows into a burial. *But wait a minute—what's all this extra stuff?* If it's something too big to move, he detours around it. But if he can drag it, he hauls it up to the surface and chucks it outside.

One digger's trash is another's treasure. During my first few days in South Dakota, I spent a lot of time walking in a half-crouch through the prairie's short grass and scrub. Most of the anthills were just piles of cast-off loess, with a few little pebbles thrown in for good measure. But eventually I began to spot other objects. Looking closer, I saw that they were tiny finger bones, weathered foot bones, and—most startling of all—flashes of brilliant color: blue glass beads, used in jewelry and as currency by the traders and Plains Indians two centuries ago. Digging one foot down, directly beneath several of these anthills, we found crumbling timbers used to close the graves. Jackpot! Fanning outward from the village, we plotted what looked to be the most promising concentration of these tiny grave markers placed by the ants. We began digging lines or rows of test squares running outward from the village site, no longer side by side but separated by five feet, sometimes ranging out twenty or even thirty feet from the prior squares.

That final, frantic push nearly killed the crew. But when it was done, we knew we'd found a huge Arikara cemetery. Judging by the dozens of graves we found in our strips of test squares, we knew there must be hundreds of burials.

But we'd run out of time. Excavating them would have to wait until the following summer.

I WAS, and am still, grateful to the industrious ants of South Dakota.

Not so, the writhing rattlesnakes. In fact, if there was one thing I was dreading as the summer of 1959 approached, it was the prospect of all those damn rattlesnakes.

The prairie is an ideal habitat for snakes. It abounds in mice, rabbits, birds, and other small prey. Like the ants, the snakes find the soil easy to tunnel into. So the population density of prairie rattlers is unsettlingly high to start with. Then came the added pressure of dwindling habitat: In 1957, Oahe Lake began to fill, and the lowlands along the river began to disappear beneath the water. So, guess what? The rattlesnakes wriggled up to higher ground—the terraces where a bunch of absentminded anthropologists were crawling through the grass, leaning into graves, reaching blindly out of pits to grope for a trowel or a brush.

Prairie rattlers are fairly small, as rattlesnakes go. Unlike diamondbacks, which can grow to six feet or more, with bodies as thick as a grave-digger's wrist, prairie rattlers rarely exceed three feet in length. But they're cantankerous, aggressive little devils, with a tendency to strike first and ask questions later. I decided that was a pretty good policy for us as well.

As a scientist, I understand that rattlesnakes fill an important ecological niche: They're a vital link in the food chain, the single most important predator keeping the prairie from being overrun by mice and other rodents. I grasp this thoroughly on an intellectual level. On an instinctual, emotional level, though, I'm terrified of the durned things. I probably shouldn't admit this, but I've always believed that the only good rattlesnake is a dead rattlesnake. When I'm confronted by a live one, my position tends to be, "This prairie isn't big enough for the two of us." Soon I developed a reputation as the fastest shovel in the West.

One of the morning rituals for an anthropology crew is to sharpen its shovels. A sharp shovel bites through soil a lot quicker than a dull one. It bites through snake a lot quicker too. Every morning we'd pass around a file and sharpen our shovels, smoothing out any nicks left by rocks, then honing the edge to razor keenness. The test of a truly sharp shovel is this: Will it shave the hair off your forearm? I didn't always take the time to lather up and shave my face, but every single morning my forearm was as bare and smooth as a baby's bottom. If I'd put a notch into the handle of my shovel for every prairie rattler it dispatched, eventually I'd have had all notch and no handle.

Snake-lovers will be appalled by my take-no-prisoners policy, but it's important to put it in perspective. First, with the reservoir rising and habitat being lost, there were far too many rattlesnakes for the remaining habitat to support anyhow. Second—and much more important to me—I had been given responsibility for the safety of the anthropology students working with me. All told, I spent fourteen summers excavating in South Dakota, a period that spanned my transition from Ph.D. student in Philadelphia to visiting instructor at the University of Nebraska to tenured professor at the University of Kansas. During that time nearly 150 students worked for me out on the plains. Quite a few prairie rattlers died from close encounters of the interspecies kind during those years. Not one of my students did.

Sadly, other students did die.

The prairie is notorious for the suddenness and violence of its weather changes, and that's especially true in summer. All that grass gives off a tremendous amount of moisture. As the sun beats down, the water vapor rises until it condenses, sometimes as puffy, cotton-candy clouds, and sometimes as black thunderheads towering four miles high.

Four students on an archaeologist's crew were returning from a remote village site by boat when a storm caught them. They had seen it coming and tried to outrun it, but a prairie storm can strike as swiftly and as mercilessly as an angry rattler. Lashed by gale-force winds and ocean-size waves, their boat capsized and all four drowned. Their boat was carrying life preservers, but—being young and feeling immortal—nobody was wearing one. Once the boat flipped, it was too late.

Sometimes students grimaced at my safety-consciousness, but I've always believed in caution, and it's always paid off: I've never been seriously hurt, and none of my students has been, either.

WE HAD RETURNED to the second terrace of the Missouri River in the summer of 1958 and excavated several dozen Arikara graves. By some archaeological standards, that would be considered highly productive. And at a site where we could return again and again for years, it would be. But at the Sully site—and every other site in the Missouri basin for 225 miles upriver—we knew we had very little time. The gates of Oahe Dam had just closed, and the waters began to rise. We had to work faster.

Ten years earlier, when I was an undergraduate student in college, I had spent my summers working in my stepdad's rock quarry, driving bulldozers and dump trucks. It was a great summer job, like being a really big kid playing with huge Tonka toys.

I've never been particularly interested in speed—fast cars hold little appeal for me—but power, well, that's another thing altogether. Give me a truck with a big diesel and a fat granny gear, and I'm a happy man.

Summers at the quarry, I took some flak because I was the boss's son. Some of it was good-natured; some of it wasn't. There was one fellow in particular—a skinny, mean guy in his forties—who seemed to go out of his way to give me a hard time. One day, as I was driving down a narrow lane between two buildings, I met him head-on, coming the other way in a flatbed.

The rules of the road at a quarry are quite specific about encounters like this: The loaded truck always has the right-of-way. My truck was carrying fifteen tons of rock; his flatbed was empty. There was no room to pass, and no room to turn around. He would have to back up.

But he didn't. I waited, and he sat there grinning at me. I honked my horn; he just grinned more widely.

I'd tried all summer to be nice to this guy, but clearly it wasn't doing any good. Something finally snapped. I jammed the gearshift into first and eased out the clutch. As the bumper of my truck kissed the front of the flatbed, his eyes got big. But he still didn't back up. So I mashed down on the gas pedal, and the big dump truck lurched forward, shoving the flatbed back.

What I didn't realize at first is that the bumper of the dump truck was nearly a foot higher than the bumper of the flatbed. This soon became evident, though, when the grill of his truck collapsed, the radiator burst, and geysers of steam shot out the front end. *Oh, damn,* I thought, but the damage was already done, so I figured I might as well keep going until I'd pushed him out of my way.

I caught a tongue-lashing from my stepdad later, but from then on, the older men at the quarry treated me with respect—and that mean son of a bitch stayed out of my way. Ever since, I've valued power above speed.

In South Dakota, though, it was speed we needed if we were to have any hope of outrunning the rising waters of the Missouri. As I fretted over the problem for the next two summers, a possible answer finally came to me: Maybe the key to speed *was* power.

On a cool morning in June 1960, a truck hauling a flatbed trailer bounced and lurched its way up to the Sully site, carrying a bulldozer and a road grader. I'd asked the National Science Foundation for a grant to rent power equipment to excavate, and—clearly with mixed feelings—they'd agreed to let me try it as an experiment.

I was banking on a particular property of the soil: The disturbed earth of an Arikara grave was darker and fluffier-looking than the denser, undisturbed loess around it, making the grave's circular outline easy for the trained eye to see. At least, that's how things worked when the top layer of soil was carefully removed by hand. Would that hold true if we used earthmoving equipment to scrape away the upper foot of topsoil?

Would we still be able to spot the burials' wood coverings and distinctive circular outlines—or would the blades and wheels of heavy machinery churn everything into one big mass of dirt and bone shards? If it did, it would be an ironic comeuppance for me, since one reason I'd come to South Dakota had been to protect the bones, not crush them.

We started in an area where the ants and our excavations had told us we'd be likely to find burials. The driver made a straight pass, eighty feet long but just two inches deep. Nothing but sod and that fine-grained loess.

Several more passes; still nothing. I was just about to call a halt, convinced that it had been a harebrained idea, when I saw it: In the wake of the scraper and the bulldozer—at that magic depth of twelve inches—was a distinct circle of darker, looser soil. I let out a whoop that would have done an Arikara warrior proud.

That summer, with the help of the power equipment, we excavated more than three hundred Arikara graves—ten times the number we'd excavated by hand the year before.

By this time, we were a regular summer colony in South Dakota. Initially we'd camped in tents at the site, but after the first couple of years we began renting a house for the crew, plus another for the Bass family, which by now included me, Ann, Charlie, and a new addition, William M. Bass IV—Billy. My crew always consisted of ten students plus one cook, who labored mightily to keep us all fed (sometimes seemingly on nothing but government-surplus peanut butter, a food I still can't eat to this day, four decades later).

The houses were sparsely furnished. Everybody slept on Army cots, slings of green or tan canvas stretched over rickety wooden frames. Early on, I noticed a problem with the cots: They kept breaking. Now, if millions of soldiers can sleep on Army cots without breaking them, a handful of students should be able to also. The problem, it soon surfaced, was sex: Two bodies in motion just put too much strain on the cots' flimsy joints. So I passed a rule, the first of my two cardinal rules for summer crews: No sex on the Army cots. The breakage stopped.

Rule number two was equally simple and much more serious: Don't get arrested—not for speeding, drinking, fighting, disturbing the peace, or so much as spitting on the sidewalk; if you do, you're out. We were under so much pressure already, from the rising waters of the river, we couldn't afford to complicate our task by antagonizing the locals. I only had to enforce rule number two one time, and I never, thank goodness, walked in on a violation of rule number one.

Even with the addition of earthmoving equipment, the work of excavation remained exhausting. We were covering a lot more ground now, but we were still moving a lot of dirt by hand. To keep the crews motivated, I'd stage games and contests—pointing out the crotch of a tree that was about to be submerged, for instance, and seeing who could hit it with the most shovelfuls of dirt. It might sound silly, but it kept morale high. The summers were hard and hot, but they were fun.

They were a scientific revelation too. As the number of graves we'd excavated mounted into the hundreds, a remarkable picture began to emerge from the prairie earth. For the first time in the history of Great Plains archaeology, we had large, documented samples of an entire tribe's skeletal remains, from birth through old age. For the Arikara, we realized, life was harsh, violent, and often very brief. We found an astonishing number of small graves containing the remains of infants and children. Tallying the statistics, we found that almost half the population died before age two; by age six, the mortality rate reached 55 percent. Then, interestingly, it plateaued: Very few deaths occurred between ages six and twelve; apparently, if you survived early childhood, you were likely to make it to puberty. Then, starting at around age sixteen, life got perilous again. The females began having babies, and the males began hunting buffalo and waging war. It was a violent, hazardous way of life.

The Arikara themselves were sedentary, but their neighbors and frequent enemies, the Sioux, were not, and often attacked. Many of the male skeletons bore deep scars from arrow wounds, especially in the pelvis and chest. We found many arrowheads embedded deep within bones. Often these wounds were fatal, but sometimes the bone healed around the flinty point, telling us that this particular warrior had lived for years with a Sioux arrowhead inside him.

Some skulls, both male and female, were crushed, reflecting the brutal efficiency of stone war clubs. And then there were the skulls bearing cut marks, usually most prominent at the hairline on the forehead, where the initial incision was made to detach the scalp. Some of these scalping victims still had flecks of flint in the skull. In a few chilling instances there was evidence of healing to the cranium: a scalping victim who had lived to tell the harrowing tale.

One thing we did not find at Sully was bullets. The village was abandoned for the last time around 1750. The whites and their weapons remained little more than a distant curiosity at that time. But in the short space of fifty years, that would change dramatically and, for the Arikara, tragically.

The Sully site was the largest of the Arikara villages. But the Leavenworth site, two hundred miles upriver, was the most poignant. It was there that the Arikara gathered, around the year 1800, to make their last stand against the Sioux, the whites, and deadly enemies they could not even see. Twelve separate Arikara bands converged, seeking safety in numbers. At a site just south of the present-day border of North Dakota, they built a pair of villages a few hundred yards apart on the first terrace of the Missouri, separated by a pleasant little stream.

It was there that Lewis and Clark encountered and scrapped with the Arikara. It was there that unscrupulous agents of fur companies waged biological warfare on them, bringing blankets from Saint Louis— blankets deliberately contaminated with smallpox, to which the Indians' unsuspecting immune systems fell easy prey. And it was there, on August 9, 1823, that Colonel Henry Leavenworth and a force of nearly three hundred U.S. Army soldiers, Missouri militiamen, and Sioux warriors attacked the villages with rifles, bows, clubs, and gunboats. During the night of August 14, the remaining Arikara slipped away from their battered villages.

. . .

BY THE SUMMER OF 1965, the water level in Lake Oahe had risen to nearly 1,525 feet above sea level—more than 100 feet above the river's natural level—and the two Arikara villages at Leavenworth had disappeared beneath the water. Fortunately for us, the two main cemeteries lay one terrace above the villages, nearly 50 feet higher. So we still had time to excavate, though the pressure was relentless.

In July of 1966, however, the water was obviously catching up with us, filling some of the burial pits even as we were excavating them (giving new meaning to the phrase *watery grave*). By that time, we had found and excavated nearly three hundred Arikara graves at the Leavenworth site. We kept working, moving uphill just ahead of the water. But then the finds ceased. We cut long swaths with the power equipment, ranging farther and farther from the main cemetery areas; we even resorted to the old-fashioned technique, digging by hand. But we found nothing more. On July 18, 1966, we abandoned the Leavenworth site to the river, just as the Arikara had done 143 years before.

Years later an Indian activist would refer to me in a newspaper interview as "Indian grave-robber number one," and I suppose it's true. Over the course of fourteen summers, I excavated somewhere between four and five thousand Indian burials on the Great Plains; as far as I know, that's more than anyone else in the world.

And yet, I never had a single clash with Native Americans during those fourteen years. There are two explanations for that. First, my wife, Ann, a nutrition scientist, spent her summers working to improve nutrition among the Sioux Indians on South Dakota's Standing Rock Reservation. Ann wrote her Ph.D. dissertation on the high rate of diabetes among the Sioux and was regarded by them as a friend. As Ann's husband, I got the benefit of the doubt. Second, I was helping the modern Sioux settle their score with the ancient Arikara: helping them "count final coup," as they call it.

But as the 1960s drew to a close, it was clear that change was coming. Lake Oahe was filling up, and the Smithsonian River Basin Surveys were winding down. Of the hundreds of archaeological sites identified before the reservoir began filling, only a small percentage was ever excavated. There wasn't enough time, money, or manpower to do more.

But we weren't just racing the rising waters; we were also swimming against a powerful new cultural current. By the late 1960s—the era of civil rights, Vietnam, and broad social upheaval—Native Americans began reasserting their claim to their culture, their heritage, and their relics. A major clash between science and cultural values was clearly brewing. Bob Dylan's folk anthem for the sixties spoke of changing times and rising waters, and it advised, "You better start swimmin' or you'll sink like a stone." With the muddy waters of the Missouri rising around my ankles, I decided it was time to start swimming.

And at that pivotal moment, the University of Tennessee came calling. So did forensic anthropology. My career as "Indian grave-robber number one" was over. My true vocation—as a forensic scientist—was about to begin.

CHAPTER 3

Bare Bones: Forensics 101

I N A RELATIONSHIP lasting forty years, you can learn a lot about someone. But everybody takes some secrets to the grave.

I first met my longtime teaching partner at the beginning of the fall semester back in 1962. I was a freshly minted Ph.D., teaching at the University of Kansas in Lawrence in between my summer excavations in South Dakota; my partner-to-be was a not-so-fresh corpse lately pulled from a roadside near the Missouri River outside Leavenworth. The body, found by three dove hunters and one bird dog, was down in the floodplain—what the locals call "the bottoms"—where the soil, deposited by occasional floods, was soft and sandy. When the murder occurred it was summertime, and the diggin' was easy.

As a forensic anthropologist, I tend to see bodies that are long past their prime—bodies that are bloated, blasted, burned, buggy, rotted, sawed, gnawed, liquefied, mummified, or dismembered. Some are even skeletonized, reduced to bare bones—bare, but brimming with data.

Flesh decays; bone endures. Flesh forgets and forgives ancient injuries; bone heals, but it always remembers: a childhood fall, a barroom brawl; the smash of a pistol butt to the temple, the quick sting of a blade between the ribs. The bones capture such moments, preserve a record of them, and reveal them to anyone with eyes trained to see the rich visual record, to hear the faint whispers rising from the dead.

I was in the morgue at University of Tennessee Medical Center recently, and laid out on a metal tray there I saw a heartbreaking sight: the skeleton of an infant, just three months old, battered beyond anything I've ever seen. An arm and a leg were broken; so was nearly every tiny rib. The most horrifying part was this: In addition to the fresh, perimortem breaks—those occurring around the time of death—there were numerous other fractures in various stages of healing. This poor child was abused almost from the moment he was born, yet his broken little body kept trying to mend itself. Given half a chance, he would have recovered, because the body's resilience can be incredible. So, too, is the depth of cruelty in some people. It did my heart good, in a sad sort of way, to read that his mother was later charged with murder and is now awaiting trial.

The adult victim I examined that day back in 1962—the one that would become my teaching partner—was not reduced to bare bones. The examination would have been much more pleasant if that had been the case. The remains arrived in a reeking cardboard box tied with twine to the trunk lid of a black sedan. The two Kansas Bureau of Investigation (KBI) agents who had tied it there didn't want to stink up their trunk. They didn't want to stink up their hands, either: "I'm not touching it," said one of them. "You'll have to go get it yourself." So I went out to the parking lot, cut the twine, and carried the box over to the side yard of the university's Museum of Natural History, which housed my office. Setting the box on the grass, I lifted out a plastic bag, untied its neck, and extracted the remains, piece by rotting piece.

A handful of my braver anthropology students gathered around. The fall term had started only hours before—it was the day after Labor Day—and already things were getting lively. Despite the gruesomeness of it all, studying a freshly unearthed murder victim was a unique learning opportunity, one that few anthropology students—and not all that many professors—ever receive.

When you examine a body in a forensic case, I told the students, the ultimate goal is to make a positive identification. If possible, you also want to determine the cause of death (technically, only medical examiners can determine cause of death; we anthropologists call things like stab wounds and gunshots "manner of death").

But before you can tell who someone was and how they died—and you won't always be able to tell—you start with the Big Four: sex, race, age, and stature.

Whenever I examine human remains, I start by laying out the body or the bones face up, in anatomical order. In this case, that didn't take long: The KBI had brought me just three pieces—a femur, a mandible (lower jaw), and a skull. Back in 1962, anthropologists were seldom brought to crime scenes to help excavate or recover remains; instead, police did the excavations as best they could (sometimes carefully, often clumsily), then brought over the skull, as in this case, or maybe a broken bone or a cut rib, and asked about whatever was puzzling them. It was rather like asking a mechanic to diagnose your car's backfiring by taking him just your carburetor or alternator instead of letting him inspect the whole car, but that's how things were done in those days. Fortunately, over the years I developed close working relationships with police, so increasingly I was called to crime scenes to recover remains as soon as they were found.

As the students leaned in for a closer look—some of them holding their breath against the smell—we studied the femur, which still had quite a lot of tissue on it. From the angle of the femoral head (the "ball" that fits into the hip joint's socket) and the lower articulating surface, where the femur joins the tibia to form the knee, I could tell we were holding the right femur. I laid it on the grass, adjoining an imaginary hipbone. Somewhere in between I mentally pictured a pelvis, a spinal column, two arms, and the rib cage. At the top of the imaginary spine I laid the head and mandible.

The face was gone. Leering up at us from the grass was a greasy, stained skull with rotting patches of skin and muscle at the sides and back of the head. For a bone man like me (this was years before the term *forensic anthropologist* was coined), the absence of flesh on the face would actually simplify our task.

Here's why: A corpse's skin can be deceptive. If a body is bloated, the facial tissues can swell, making it more difficult to discern the person's gender. If the genitals are missing—because of dismemberment, decomposition, putrefaction, or animal feeding—or the soft tissue is badly decomposed, the shapes of the bones themselves will offer the most reliable information.

This particular skull was small, which immediately suggested either a child or a woman. The mouth was narrow and the chin was pointed—additional features characteristic of a woman. The forehead was gracile—smooth or streamlined, particularly the forehead and the ridge above the eyebrows: a textbook example of a woman's skull, I told the students.

"You've probably seen cartoons of big, hulking Neanderthal cavemen," I said. "The men have these massive brow ridges, so that when another caveman clubs them with a woolly mammoth femur, it doesn't hurt." They laughed at that; over the years I've found that humor helps students learn, so I always look for opportunities to throw in jokes that will reinforce what I'm explaining. "I'm not saying that we men have failed to evolve in the last twenty thousand years, but a modern male skull looks a whole lot more like a Neanderthal's than a modern female skull does."

Holding the skull up so they could see it better (and smell it better, too, unfortunately), I showed them the brow ridge above the eyes. Lacking the massive ridge of the male, a woman's skull has sharp edges where the eye orbits, or sockets, are set beneath the forehead. Finally, turning the head around, I showed them the base of the skull—the occipital bone—where men have a bony bump called the external occipital protuberance. This skull didn't; clearly this was not a manly man.

"But how can you tell for sure," I asked the students, "whether this was an adult woman or a twelve-year-old boy?"

One of the students hazarded a guess: "The teeth?"

"That's right," I said, "the teeth."

Our mystery victim had a full set of teeth—thirty, including the upper pair of third molars, or wisdom teeth, though not the lower pair. One evolutionary change we humans are undergoing as we've given up gnawing on animal bones is the gradual loss of our third molars. Some people's wisdom teeth never erupt; they're like a seed that never germinates. So, finding a skull in which the third molars *haven't* erupted, I explained, doesn't necessarily mean that the person was not yet an adult. However, if the third molars *have* erupted, I stressed, it's virtually certain that the individual was eighteen or older. In this case, then, I was pretty sure we were looking at an adult woman.

The best way to confirm that, I added, would be to examine the pelvis. It was a shame we didn't have it.

The adult pelvis is a complexly engineered structure resulting from the union of three rugged bones: the sacrum, at the base of the spinal column, and the two hipbones, the right innominate bone and the left innominate. (The term *innominate*, which translates as "nameless" or "unnamable" bone, is a comment on its odd shape: Viewed from the front, the hipbones flare out like the ears of an angry elephant; underneath those flaring, bony ears are two knobs pierced by openings like empty eye sockets; in front, two prongs of bone converge like tusks grown badly awry.)

The sacrum acts as a weight distributor, splitting the weight from a single column, the spine, into two columns, the legs, by way of the right innominate and the left innominate. But the innominate itself is a complicated structure, somewhat analogous to the cranium, which is also formed by the fusion of multiple bones.

Before puberty, each innominate bone consists of three separate bones: the ilium, the ischium, and the pubis. The ilium is the highest, broadest part of the hipbone; its crest is what flares out like elephant ears just below the waist. The ischium is the bony structure you can feel yourself sitting on, if you wiggle your butt on a hard, wooden chair. (Some of us have a hard time feeling anything bony inside that big blob of fatty tissue, but it's there nonetheless.) The pubis is the bone that spans the front of the abdomen, about four inches or so below the navel.

At puberty the pelvis gets interesting in many ways, including skeletally. To allow passage of a baby's head during childbirth, the female's hipbone gradually broadens and the pubic bone gets longer, angling farther forward to form more of an arch for the birth canal.

Because the male's pelvis is markedly narrower, his femurs hang roughly straight down below his hipbones. In an adult female the femurs incline slightly inward beneath the hips. Not surprisingly, this difference in pelvic and femur geometry translates into some scientifically observable and aesthetically pleasing differences in the way men and women sit, stand, and walk.

In the case of our recently unearthed murder victim, then, having the pelvis would have easily confirmed that the skull was a woman's.

The pelvis would also have told us more about our victim's age. Like the sutures in the skull, the joint at the body's midline where the left pubis meets the right pubis—called the pubic symphysis—is an excellent yardstick for measuring age. From late adolescence through about age fifty or so, the bony face of the pubic symphysis undergoes a gradual, consistent set of changes, which were first studied and cataloged more than eighty years ago: Corrugated or bumpy during a female's late teens, the pubic symphysis smoothes out during the twenties and thirties; by age forty or so, its face begins to erode and acquires a porous, spongy look. Considered along with other skeletal features such as teeth, cranial sutures, and the degree to which the ends of clavicles (collarbones) have fused to their shafts, the pubic symphysis allows an anthropologist to estimate age with remarkable accuracy—often within a year or two of the victim's actual age.

For determining race, though, we had everything we needed in the skull. I directed the students' attention again to the woman's mouth. Her teeth jutted sharply forward; so did her jawbones in the region where the teeth were rooted. It's a trait called prognathism (from ancient Greek, meaning literally "forward jaw"); even novice anthropologists can readily recognize it as one of the hallmarks of Negroid skulls.

There's an easy test for prognathism, I told them, and I demonstrated with the skull in my hand. Take a pencil and press one end between your upper lip and the base of your nose. Holding that end in place as a pivot point, swivel the pencil downward. If it contacts the lips and teeth but can't touch the chin, your skull is prognathic and probably Negroid; if it can touch both the base of the nasal opening and the tip of the chin, your skull is orthognathic (flat) and probably Caucasoid.

Our skull passed the pencil test for prognathism with flying colors; her jaw morphology was a textbook example of Negroid structure. The teeth themselves were further confirmation: The tops of her molars were rugged and bumpy—crenulated, anthropologists call it—unlike the smoother cusps of Caucasoid teeth.

A word about race: In recent years, the very concept of distinct races has come under attack. Race is merely a cultural construct, says one recent school of thought, not an objective physical or genetic feature. On the one hand, it can be useful to question and rethink our notions of what race means; on the other hand, I've examined tens of thousands of skulls over the course of nearly half a century, and their features—visually distinct, numerically measurable, and statistically graphable—correspond quite consistently to three main groupings: Negroid, Caucasoid, and Mongoloid. (Anthropologically, *Mongoloid* refers to Asian, Eskimo, or Native American ancestry, not Down's syndrome.) As the world's peoples increasingly mingle, traditional racial distinctions and labels may eventually blur and even disappear, but in the meantime I'll hang on to them, because they help me identify the dead and they help police solve murders.

By now, the students had absorbed enough knowledge and enough odor for one hot afternoon. I returned the skull and femur to their plastic bag, closed the box, and took it to my car. Unlike the KBI agents, I put the box in the trunk. I wasn't quite willing to put the remains in the passenger compartment, but I was willing to bring them into our kitchen and simmer them on Ann's stove.

To refine my estimate of age and to gauge the woman's stature, I needed to remove the remaining tissue from the bones. Short of leaving the skull and femur outdoors and allowing insects and scavengers to pick the bones clean—a slow process, and one that could mean losing the femur or mandible to some scavenging buzzard or coyote—the only good way to clean the bones was to simmer them in a covered steam vat for the better part of a day, then scrub off the softened tissue with a toothbrush (not my own personal one, mind you).

Ann was a nutrition scientist; she took her cooking, and her kitchen, very seriously. Needless to say, she wasn't thrilled when she arrived home to the stench of cooking flesh and found a decaying human skull and femur simmering in her eight-quart kettle. She'd walked in on this more than once: Part of the University of Kansas anthropology department, including my office, was housed in the Museum of Natural History; it was a grand old building, but it was built to house old, dry bones, not process fresh, tissue-covered ones. As a scientist herself, Ann realized I had to get the work done any way I could. Marriage survives on compromise, and we had hammered out some unorthodox but workable ones: She tolerated my occasional use of her stove for processing remains, but her pots and pans were strictly off limits—I had to provide my own.

It's true what they say: A watched pot never boils. However, an unattended one—at least if it's filled with human bones and decomposing flesh—swiftly bubbles over. I left my post at the stove just long enough to go to the bathroom; when I returned, a froth of water, brain soup, and other foul-smelling components was pouring over the rim and seeping into every recess of Ann's stove. It would never be the same. From that day on, moments after a burner or the oven was switched on, that same foul odor would curl upward and fill the kitchen. Exercising my incredible powers of scientific deduction, I swiftly deduced that daily reminders of my lapse at the stove might not be conducive to marital harmony, so in very short order, Ann was the proud owner of a new kitchen stove.

Meanwhile, I had scrubbed the bones and set them out in the early-September sunshine to dry. Scrubbed clean of all its soft tissue, the skull gleamed with a smooth, ivorylike sheen—another characteristic of Negroid skulls, whose bone is denser than Caucasoid skulls. The mouth's prognathism was even more pronounced, now that there was no tissue altering the skull's contours. The nasal opening was broad, with vertical "guttering" in the upper jaw—distinctly different from the horizontal sill or "dam" at the base of a Caucasian's nasal opening. (The broad, unimpeded nasal opening in the Negroid skull evolved to promote rapid air exchange and cooling in hot climates; the narrower opening and nasal dam in Caucasoids evolved to keep cold European air from flowing too rapidly into the lungs.)

So by now I knew these bones were from a Negro female and I knew she was an adult. But was she eighteen or was she eighty? To find out, I looked to the cranial sutures.

Most people think of the cranium as a single dome of bone, and if you run your hands over the top of your head, it certainly *feels* like one piece. In reality, though, the cranial vault is a complex assembly of seven separate bones: the frontal bone, or forehead; a pair of parietal bones, which forms the skull's upper sides and rear; the temporal bones, low on either side; the sphenoid, which forms the floor and part of the sides, and the occipital bone, the skull's heavy back and base, which rests atop the first cervical vertebra and channels the spinal cord into the neck. (For a labeled diagram of the skull, see Appendix I, "Bones of the Human Skeleton.")

The joints where the cranium's seven bones meet are called sutures. The name refers to their appearance: They have a serrated or zigzag look, like the ragged stitches holding together Dr. Frankenstein's monster. When we're born, the joints are actually formed of cartilage, but as we age, the cartilage ossifies (turns to bone) and the sutures smooth over, all but disappearing by old age in many cases.

This woman's coronal cranial suture—the one running across the top of her head—had begun to fuse; that meant she must have been at least twenty-eight, because generally that joint is one of the last to fuse. But the fact that it was only partially fused indicated that she was probably not far past thirty—probably thirty-four at the oldest, I estimated.

So far so good: I knew three of the Big Four—sex, race, and age. That left only stature. For centuries, artists and scientists have noticed that although people's height or stature can vary enormously, their *proportions*—the ratio of leg length to total stature, for instance—are all pretty much the same. There's a famous illustration in Leonardo da Vinci's notebooks showing a nude man drawn within a circle and a square; he's drawn with four arms (one pair stretched out to the side horizontally, the other pair elevated so the fingertips are at the same height as the top of his head) and four legs (one pair with the feet together, the other pair with the feet spread several feet apart). In his trademark mirror-image script below the illustration, Leonardo adds these observations on human proportion developed by the architect Vitruvius: "The length of a man's outspread arms is equal to his height. . . . The greatest width of the shoulders contains in itself the fourth part of man. From the elbow to the tip of the hand will be the fifth part of a man; and from the elbow to the angle of the armpit will be the eighth part of man. The whole hand will be the tenth part of the man."*

In the 1950s anthropologist Mildred Trotter and statistician Goldine

* Reprinted by permission of the publishers and the Trustees of the Loeb Classical Library from VITRUVIUS—ON ARCHITECTURE, Vol. I, Loeb Classical Library Vol. L251, translated by Frank Granger, Cambridge, Mass.: Harvard University Press, 1931. The Loeb Classical Library® is a registered trademark of the President and Fellows of Harvard College.

Gleser took that age-old notion of proportionality and conducted extensive skeletal research to refine its accuracy. After measuring hundreds of skeletons, Trotter and Gleser devised formulas that could extrapolate stature from the length of any of the body's so-called long bones—the bones of the arms (humerus, radius, or ulna) or legs (femur, tibia, or fibula). The best results come from measuring the femur, the thighbone; that's probably why the KBI brought me a femur.

Placing the bone on an osteometric board—a sliding wooden gizmo that resembles a pair of bookends joined by a yardstick—I measured its length at 47.2 centimeters. Then I plugged that number into Trotter and Gleser's formula for Negroid females: $(47.2 \times 2.28) + 59.76$. The resulting number, 167.38, was her stature in centimeters. Translating from metric measurements to English units told me that the woman was about five feet six inches, give or take an inch.

So now I knew all four: sex, female; race, black; age, thirty to thirty-four; height, five feet six inches. The next question would be harder to answer definitively: Who was she? Normally, when a skull comes in with a full set of teeth, there's a reasonably good chance of making a positive identification. The trick is to match preexisting dental X rays with the corpse's fillings or bridgework or other unique features in the shape, structure, or arrangement of the teeth. Of course, to do that, you've got to lay your hands on the dental X rays of missing persons who match the age range, sex, and race of your corpse. That isn't always possible, but you'd be surprised how often a dentist is able to provide the records needed to cinch an identification.

In this case, though, there was a problem: This woman's teeth showed no signs of any dental work. Lord knows, she could have *used* some dental work: She had large cavities in two of her lower teeth and in five of her upper teeth, and smaller cavities in most of her other teeth. Worse, one of her upper wisdom teeth had abscessed. The lack of dental care meant she was probably poor; the fact that she'd managed to hang on to her teeth so far, and had been able to withstand the pain of an abscess, suggested that she was one tough cookie. One other feature of her dentition was striking: When I fitted her mandible to her skull, I couldn't quite get her lower jaw to line up beneath her upper jaw; the mandible skewed about a quarter-inch to the right, giving her a slight but distinctive crossbite that would have shown up whenever she flashed a big smile.

Lacking dental work, dental records, or photographs, I couldn't make a positive identification of the body. However, I could make a presumptive, or probable, identification. A woman from Atchison, Kansas— a small town about twenty miles from where the body had been found—had been reported missing on August 10, some three weeks earlier. Her name was Mary Louise Downing; she was a black female, age thirty-two, height five feet six inches. There was no ironclad, 100 percent guarantee that the skull and femur I had were hers, but there was certainly nothing in my examination that cast any doubt on it, either. In fact, I'd have been willing to bet the price of that new kitchen stove that this was Mary Louise.

On Saturday, September 8, I typed up my report and mailed it to the lead KBI agent investigating the case, along with a copy to the KBI's director in Topeka. Single-spaced, the report didn't even fill two pages.

In the end, there wasn't really much I could tell the KBI about her beyond her sex, race, age, stature, and poor dental health. The skull and femur revealed nothing about her manner of death. But apparently the KBI had more to work with than I did, and after my examination and report, they were confident that Mary Louise Downing had indeed been found. From the fact that she'd been hidden in a remote stretch of the river bottoms, they assumed she'd been murdered.

But that was it. Who had killed her, and why, and where, and when— those were secrets that only two people possessed, the killer and Mary Louise, and neither one was talking.

After mailing off the report, I took one more look at her skull. Piercing her cheekbones and lower jaw, about an inch and a half on either side of the skull's midline, were four neat, tiny holes where the craniofacial nerves had emerged from her brain. Thin bundles of electrochemical fibers, they would have translated this woman's inner sadness into outward frowns, her purest happiness into the slightly sideways smile that her crossbite would have given her. She had been someone's daughter, someone's wife, someone's mother. Now she was reduced to a case—one that would never be solved.

Her disappearance that day in August hadn't merited a mention in the local newspaper; the discovery of her body in early September had rated only two column inches. In death, as in life, Mary Louise seemed destined to fall through the cracks, unnoticed, uncared for, insignificant.

And yet . . . And yet . . . We've spent forty years together now, Mary Louise and I. She's been in almost every classroom I've set foot in; she's traveled with me to seminars and conferences, all over the United States: the FBI Academy at Quantico, Virginia; the Bureau of Alcohol, Tobacco and Firearms trainings in half a dozen states; the U.S. Army Central Identification Laboratory in Honolulu, Hawaii. In life Mary Louise probably never traveled far from Atchison or accomplished much that would look noteworthy. But in death she's journeyed halfway around the world, educated thousands of students, and helped train hundreds of forensic anthropologists, homicide investigators, crime lab technicians, and medical examiners.

Mary Louise's murder will probably never be solved. But, thanks to her, other murders will be—and probably already have been. To me, that makes her a remarkable woman, and a forensic hero.

No bones about it.

CHAPTER 4

The Unsavory Uncle

A SHERIFF'S DEPUTY appeared in the doorway of my office in the Museum of Natural History at the University of Kansas in Lawrence in December of 1970. Six months later and the deputy wouldn't have found me in Kansas. I had already accepted a new job at the University of Tennessee in Knoxville, and we planned to move the following May.

The deputy caught me at the desk where I'd spent falls, winters, and springs for the past decade. During that time, the University of Kansas had built one of the best physical-anthropology programs in the nation. With three young, innovative physical anthropologists on the faculty, we were becoming widely known for our forensic expertise. By now I'd worked dozens of forensic cases for various law enforcement agencies, ranging from tiny sheriffs' offices to the Kansas Bureau of Investigation, whose associate director, Harold Nye, had become a good friend of mine.

Harold was something of a celebrity in law enforcement circles by this time. He played the key role in tracking down the two former convicts who murdered a family of four in west Kansas in 1959. The case—the murder of the Clutter family, and the KBI's pursuit of their killers—sparked one of the all-time classics of crime writing, Truman Capote's *In Cold Blood*, which came out in 1965.

Capote recounted how Nye fought a persistent case of the flu during the six weeks it took to catch the killers, ex-convicts Dick Hickock and Perry Smith. Despite fevers, Harold worked tirelessly as part of the team of four KBI agents assigned to the case. He followed Perry Smith's trail to a cheap rooming house in Las Vegas where he had stayed shortly before the murders; more important, he learned from the manager that she was expecting Smith to return to claim a box of possessions he'd stored there. In Mexico City—one of many places the killers traveled after the crime—Harold managed to find a pair of binoculars and a transistor radio they'd stolen from the Clutter house and pawned for a few dollars. Those were important pieces of evidence at the trial, because they helped prove the men had been at the house.

Harold also snagged another key piece of evidence at the murder scene itself. Two distinctive sets of boot prints, too faint for the human eye to notice, showed up in Harold's photos of the Clutters' basement floor. When the killers were arrested, their boots matched those prints exactly. Thanks to meticulous casework by Harold and the other KBI agents, the two men were convicted of first-degree murder and hanged.

Harold didn't much like Truman Capote's account of the case; Nye thought it took too many liberties with facts. He also didn't think much of Capote himself: When Harold went to Capote's hotel room to do an interview, Capote answered the door wearing a lacy negligee. That must have given straight-arrow Harold quite a jolt, but he kept it to himself until years later, when he told the story to the writer George Plimpton, who was doing a biography of Capote.

Although neither of us knew it at the time, Harold would eventually help inspire the creation of the Body Farm. One spring day back in 1964,

he called with an unusual question: Could I examine a skeleton and estimate the time since death? This particular skeleton, it turned out, belonged to a cow; occasionally cattle rustlers or vandals leave dead,
mutilated cattle out on the prairie. And since there are more cows than
people in Kansas, the KBI spent a fair amount of time investigating cattle rustling. In this instance, rather than rustling the cattle, the thieves
simply killed and field-dressed the cows, taking the meat and leaving the
bones.

A few days after his call, after double-checking with the university's
paleontologist, I sent Harold a letter. "We do not know of any method
by which you could tell the length of time since the cow has been
killed," I wrote. "I can tell you the age of the cow at death; however, I
cannot tell you how long it has been since the cow was killed."

But his request had set me to thinking. "I do have a suggestion." I
continued:

> As you can imagine, there has been no work done on this that we
> are aware of. If you have some interested farmer who would be
> willing to kill a cow and let it lie, we could run an experiment on
> how long it would take for the flesh to decay and begin to build
> up some information in this area. However, the rate of decom-
> position is not the same in the summer as it is in the winter and
> I am afraid that we would have to sacrifice at least two or more
> cows before we could get complete data . . .

Harold never followed up on my suggestion; I guess it was the scientific equivalent of Truman Capote answering the door in a woman's
negligee—maybe just a little too unusual for his tastes. But then again,
I didn't rush to pursue it, either. The fact is, I forgot all about it for
nearly forty years; recently I came across that letter in a dusty file,
tucked behind a crackly X ray.

But even though I'd filed and forgotten that brief scientific suggestion,
somewhere in my subconscious, a seed had been planted—a seed that

would germinate some fifteen years later and bear scientific fruit, arising not from dead cows but from human corpses: corpses at the Body Farm.

But I'm getting ahead of my story. The Body Farm was still far in the future; this was December of 1970, and a detective from the nearby town of Olathe, twenty-five miles southeast of Lawrence, entered my office carrying a cardboard evidence box. Inside was a small, sad set of skeletal remains. I could tell at a glance they were the bones of a small child, probably no more than two or three years old. The sheriff's deputy, Detective Jerry Foote, told me they had been found a week before by quail hunters out on the prairie. Most of the bones were missing, which I suspected was due to scattering or consumption by animals; fortunately, the skull was relatively complete, except for the absence of most of the teeth.

I did an initial exam there in my office, explaining what I observed to Detective Foote. I'd learned early on that most police officers are eager to learn all they can about investigative techniques; they appreciate hearing what I have to say as I examine a body or a skeleton, even at the early stages.

As I studied this small skull, I could tell by how weathered it was that it had been outdoors for months. In addition, I noticed that the left side was bleached nearly white, suggesting that it had lain on its right side, exposing the left to the brunt of the sun and the rain. On the right side I found a few strands of fine blond hair stuck to the forehead, as well as some at the base of the skull and the cervical vertebrae. The hair confirmed what I had thought immediately from the shape of the skull: this child was probably Caucasian.

Most of the teeth had fallen out, but it was clear that the child had a nearly complete set, including the first molars, which were still attached; that told me the child was probably at least twenty-four months old. The roots of the canines, however, had not yet formed completely, which meant the age was less than thirty-six months. Three years old: For most children, it's the age of nursery rhymes, stuffed animals, hide-and-seek, crayons. For this child, it was the age of death, and possibly murder.

Was it a boy or a girl? By adolescence the sex of an unidentified skeleton can be determined fairly easily, mainly from the pelvis: Females have a wider pelvic structure and a markedly longer pubic bone, to allow for childbirth. In early childhood, though, there's virtually no difference between a male's pelvis and a female's. At any specific age, young girls tend to be a bit smaller than boys, but unless you know the age for sure—which means you probably know the identity already—you have no basis for gauging the sex.

Detective Foote told me he was fairly sure he knew the child's identity. Eight months before, Lisa Elaine Silvers, aged two and a half, had been reported missing. Her twenty-one-year-old uncle, Gerald Silvers, was baby-sitting Lisa and her baby sister on April 22, 1970, while her parents went to a movie. Gerald fell asleep, he told police, and when he awoke from his nap, Lisa was gone. A search by police and neighbors failed to find any trace of the child.

After his questioning, Gerald left Kansas and went to California—on short notice and in a police car. While doing a routine background check after Lisa disappeared, Detective Foote learned that young Gerald was wanted for second-degree robbery and hit-and-run in the Golden Gate State—not the sort of uncle I'd want baby-sitting for *my* children. But that didn't necessarily mean he was a murderer. In fact, from the contents of the box sitting on my desk, we couldn't even say for certain that these bones were Lisa's. Not only was it impossible to determine the sex of the skeleton, there were no healed injuries that could be corroborated by X rays from Lisa's medical records. Furthermore, there were no dental records; she hadn't even lived long enough to make her first visit to the dentist. I had half a hundred bones sitting right in front of me, but I didn't have a single thing I could latch on to. I wrote up my brief findings on the spot for Detective Foote and wished him good luck with the case.

A few months later Foote seemed to have had excellent luck: Two of Gerald Silvers's fellow inmates in California snitched on him, saying he'd bragged about raping and killing his niece. Gerald was indicted by a Kansas grand jury and brought back to Olathe to stand trial. But as the initial hearing approached, Detective Foote called me in a panic. Because we hadn't been able to identify the body positively as Lisa's, it would be easy for Gerald's attorney to attack the prosecution's case. There was a body, all right, but there was no particular reason for a jury to believe that it was Lisa's or that she'd been raped and murdered by her uncle.

Foote was practically pleading: Wasn't there anything else we could do to get a positive identification? "Do you have a picture of Lisa?" I asked, hoping it might reveal some distinctive feature in her facial structure that we could correlate to her skull. Yes, he did; he agreed to send it to me.

When the envelope arrived, I tore it open. The picture showed a pretty, blond, happy little girl, smiling proudly at the camera. The teeth caught my eye: Somehow, though I couldn't say why, I saw a glimmer of hope in that bright smile. I put in a call to Detective Foote.

"Tell me more about where the body was found," I said. The quail hunters who found it had been wading along a narrow, shallow stream running through a pasture, Foote told me, about ten miles outside Olathe. "We need the rest of her teeth," I said, "not just the molars."

Detective Foote sounded doubtful. They'd searched for hours, he said, to come up with this much of the skeleton. He didn't see how they could have missed anything. By this point in my anthropology career, though, I had excavated several thousand skeletons, and I'd gotten pretty good at rounding up bones and teeth. Most of those skeletons came from undisturbed Indian graves, true, but a sizable minority—several hundred, at least—had been scattered in some way: by animals, by storms or erosion, or by human intrusion. In those cases there tended to be a pattern to the scattering, and I hoped it would hold true in this one. "Those teeth will be where that body was found," I told him. "Let's go back and find them."

. . .

IT WAS MID-APRIL BY NOW, five months after the quail hunters had stumbled across the small skull in the stream. As we bumped across the prairie and stopped by the embankment, I hoped nothing had disturbed the streambed since the fall. A herd of cows stomping around in the mud could make it virtually impossible to find anything more. Fortunately, there were no signs of cattle, and we'd had a fairly warm, dry spring, so the stream was only a few inches deep. I felt my optimism returning.

It doesn't take a rocket scientist to figure out that bones in a creek will tend to wash downstream. The tricky part is figuring out how far downstream. Generally the smaller, lighter bones get carried farther than the skull or long bones. Complicating the picture slightly is the fact that the farther downstream a bone gets carried, the farther to either side it can drift as well. If you plot it on a diagram, the scatter pattern tends to look like a skinny teardrop, with the sharp end farthest upstream. The larger the stream and the faster the current, the bigger that teardrop area gets.

I went about fifteen yards downstream from the point where the skull and most of the bones had been found, so I could work my way upstream against the current. By starting beyond the boundary of the expected scatter, I'd be less likely to step on a bone and break it or mash it deeper into the mud. Working upstream also meant that the mud I stirred up as I walked and felt around in the streambed would get washed away from the direction I was heading, rather than into it. It's simple once you think about it, but you'd be surprised how often untrained searchers wade around at random, muddying up the water in more ways than one.

About ten yards downstream from the skull's location, I started feeling little pebbles in the silt. Except they weren't pebbles: They were tiny bones—hand bones and foot bones and vertebrae. And teeth—fourteen in all!—with only two, a pair of lower incisors, remaining lost. I felt like I'd hit the mother lode. As I headed back to my office in Lawrence, I hoped that somewhere in these bones and teeth I'd find something that said, unequivocally, "I am—I *was*—Lisa Silvers."

At the very least, I was sure the teeth could help refine my estimate of the dead child's age. A group of dental researchers at Harvard had carefully charted the stages of formation of several types of deciduous teeth ("baby teeth"). I x-rayed a lower canine, a lower first molar, and a lower second molar; comparing these X rays with those from the Harvard study, I got an estimate of 2.1 years. By a different study's yardstick, the first permanent lower molar suggested an age of 2.9 to 3.9 years. Yet another dental yardstick indicated an age of 2.5 to 3 years.

Of course, the real clincher in forensic dentistry is finding dental work that can be matched with dental records. Unfortunately, since Lisa had never been to the dentist, we had no dental records. On the other hand, since none of these teeth had fillings, they didn't rule out the possibility that this was Lisa.

By this time I'd stared at those teeth for hours. I could close my eyes and still see their outlines. And even though I was pretty sure there was no scientific stone I'd left unturned, I kept staring at them, turning them over and over in my hands and in my mind. It was the incisors I kept coming back to. There was something about them I was almost noticing but not quite. Maybe I was looking too closely. If you've ever stargazed, you've probably noticed that your peripheral vision can detect fainter stars than your central vision. So the trick, if you're hunting a faint star, is to look slightly away from where you think it is.

In this case, there was a way I needed to refocus or shift my vision so that I'd see what I hadn't been able to spot dead-on. So I stepped back a bit; instead of scrutinizing the teeth individually, I inserted them into their sockets in the jaws of the skull, and I looked back and forth from the skull to the photo of Lisa Silvers, alive and smiling. And that's when I saw two things I'd missed before. First, there was a slight space between the two upper central incisors—the "two front teeth," as the old song calls them. I noticed it when I fitted the teeth into their sockets, and there it was in the photo as well.

Second—and far more striking, now that I had the teeth in place—there was a slight notch at one corner of each of the four upper incisors. The teeth weren't chipped; they were formed that way. It was a genetic anomaly, and it just might hold the key to identifying this body. As I swung my gaze back to the photo, I felt a tingle of excitement. I called Detective Foote. "We have a positive identification of Lisa Silvers," I told him.

THAT WAS IN APRIL. In the two months since then, a lot of water had flowed under a lot of bridges.

For me, the biggest change was my move to Tennessee at the end of May. My years at Kansas had been a period of tremendous growth. My summers in the field were intense but exciting; the academic year brought the combined pleasures of forensic cases for the police and KBI and the daily thrill of classroom teaching. Put me in front of a group—whether it's freshman undergraduates, an anthropology Ph.D. seminar, a class of new FBI trainees, or a bunch of senior citizens—and it's like throwing a switch inside me that releases a huge jolt of adrenaline. I move around in goofy ways to show how the skeleton works; I tell jokes, usually slightly off-color ones that tend to get me hauled onto the carpet at least once a semester. But the vast majority of students seemed to notice and appreciate my teaching style; my "Intro to Anthropology" classes at Kansas swelled to more than a thousand students every fall; to cope with the flood of students, the dean had to move us from a lecture hall to the university's main auditorium.

But there was an undercurrent of deep discord within the anthropology department. When I had arrived in Kansas in 1960, the anthropology faculty consisted solely of archaeologists and cultural anthropologists.

Then, in rapid succession, three physical anthropologists were hired. Soon the three of us were building a national reputation for our forensic work—and were also teaching the majority of students who took anthropology courses. And soon the cultural anthropologists began resenting us. The tension got so bad, all three of the physical anthropologists began job-hunting.

I was the first to jump ship. The University of Tennessee was hoping to build a national-caliber anthropology program, just as we'd begun to do at Kansas. When they offered me the chance to head it—and the chance to hire two additional faculty of my choice—it was too good to turn down.

Within a year the two other physical anthropologists had likewise left for greener pastures, or at least more collegial ones, and Kansas had lost a cadre of expertise that had taken a decade to build.

When I arrived in Knoxville on June 1, 1971, it didn't look like a dream assignment. Up until then, the handful of anthropologists were housed in the university's small archaeological museum. If we were to build the department—and start a graduate program—we'd need more space, and lots of it. The only space available had just opened up: a spooky building tucked beneath the stands of Neyland Stadium, UT's enormous shrine to Southeastern Conference college football (the third-largest stadium in the United States).

The gloomy building, added in the 1940s, had originally housed the school's football players and other athletes. Then, when it got too old and run-down for the athletes, the university built a new athletic dorm and shifted nonathletes into the rooms beneath the stands. Now that the space had gotten too old and run-down for the nonathletes, the school graciously gave it to the faculty. My faculty.

What matters, though, is not the space you're put in to work; what matters is the work you do in it. The Manhattan Project, the World War II race to develop the atomic bomb, also started out under a football stadium. Beneath the stands of Stagg Field at the University of Chicago, a team of physicists led by Enrico Fermi built a crude fission reactor, brought its uranium fuel to critical mass, and set off a chain reaction that changed the world.

We started out in Knoxville with eight offices, utterly empty except for a single telephone sitting in the floor of one office. No desks, no chairs, no shelves, no filing cabinets. The moment I arrived, we began frantically scrounging, begging, and borrowing furnishings, equipment, and supplies. We never stopped. Our growth would always outstrip our budget; by now the anthropology department has grown from its original eight rooms to 150 or so. They're even older and more run-down today than they were in June of 1971, but there's still a critical mass of anthropological expertise down there beneath the stands. The chain reaction is still going strong.

NOT LONG after Lisa Silvers had disappeared, her uncle Gerald was hauled back to Tracy, California, and sentenced to an "indeterminate" length of time in the Deuel Vocational Institute for the robbery and hit-and-run crimes he'd committed there several years before.

From the beginning, police in Kansas had been suspicious of Gerald's story. Lisa had never wandered off before, and it seemed unlikely she'd have done so in the dark, while her parents were away. Most child abductions, they also knew, involve a relative or acquaintance of the victim. As their investigation continued they grew more certain of his guilt. When two of his fellow inmates at Deuel told detectives that Gerald had admitted raping and killing the toddler, they knew they had a case.

The trial was set to begin on June 16 in Olathe, Kansas; the prosecutor, Mark Bennett, had scheduled me to testify the morning of Friday, June 18. "If you arrive by plane, I will make arrangements to have you picked up if you will advise me of the flight and time of arrival," he wrote me. I wrote back to let him know that I needed to drive, so I could retrieve a few more boxes we hadn't been able to cram into the moving van that brought our belongings to Knoxville.

I had barely had time to unpack my suitcase and start settling into my new digs in Knoxville, Tennessee, when I found myself climbing into the car for the long drive back to Kansas. As I headed west on Interstate 40 in my new "Grabber Blue" Mustang convertible—my reward to myself for landing a new job and a big raise—I had plenty of time to reflect on the sad case.

I arrived on the afternoon of the seventeenth, tired from the twelve-hour drive and nervous about how my testimony would go. I reviewed my reports and mentally practiced explaining the scientific data in language that wouldn't intimidate a jury of Kansas laypeople.

The next morning, right on schedule, I was sworn in. Mark Bennett led me through my findings, briefly going through the various methods I used to determine the age, then focusing on how the gap in the front teeth and the notches on the incisors matched the photo of Lisa exactly.

To my great relief, the defense attorney didn't challenge my identification of Lisa's body. He did, however, challenge several obvious weaknesses in the prosecution's case, as I'd expected he might: Could I determine the cause of death? No, I could not. Were there signs of violence or trauma? No, there were not. Could I tell whether Lisa had been raped? No, I could not. I knew who she was, and I knew she'd been in that stream a long time, and I knew that was a human tragedy and a damned shame, but that's all I knew.

The trial lasted a week. By the time it ended I was back in Knoxville, unpacking more household boxes and desperately scrounging around for more office furniture. Mark Bennett sent me the front-page story from *The Kansas City Star*: SILVERS ACQUITTED IN NIECE'S DEATH. The defense had attacked the credibility of the two prisoners who testified that Gerald had admitted raping and killing Lisa. Both men, defense witnesses testified, were homosexuals.

Earl Silvers, Lisa's father, praised Gerald's defense attorney after the trial. "He was very good," Earl told a local newspaper reporter. "He was always working—seven days a week, up until 9 or 10 o'clock each night." Charles Silvers, Lisa's grandfather, expressed his hope that Gerald would come home to Kansas when he finished his prison sentence at Deuel. "California is not a place to start a new life," he said.

Lisa's remains were buried not long after the trial ended. If she had lived, she would be in her mid-thirties now. She might have a child of her own. Maybe a girl with fine blond hair and a slight gap in the middle of four distinctly notched teeth at the center of a big, bright smile.

CHAPTER 5

The Case of the
Headless Corpse

I T MUST HAVE BEEN a really quiet news day; there's no other possible explanation for the explosion of media interest in my slight miscalculation.

Actually, it was a quiet couple of weeks, at least to begin with. It all started during that dependably slow week in Knoxville between Christmas and New Year's Day. The university was closed for Christmas break; most of my students had gone to visit their families. My oldest son, Charlie, who was twenty-one at the time, had come to Tennessee for the holidays from the University of Arizona, where he was a first-year graduate student in—what else?—anthropology, with an emphasis in forensics. (This was back before he realized he didn't want to live on a professor's salary all his life.)

Late on the afternoon of Thursday, December 29, 1977, I got a call from the Williamson County sheriff's office. Because I was the Tennessee state forensic anthropologist—as well as a badge-carrying consultant to the Tennessee Bureau of Investigation—law enforcement officials throughout the state had my home telephone number. Consequently the phone was apt to ring at any hour of the day or night, and the less convenient the hour, the more likely it was to be a call from someone who needed me to examine a body.

This time the someone was Detective Captain Jeff Long, calling from Franklin, a town about thirty miles south of Nashville. Franklin was small at the time—just a few thousand people—but a lot of country-music stars and Nashville doctors owned horse farms and mansions there, so it was a town of relatively affluent and well-educated people.

Two of the most affluent and best educated were Ben and Mary Griffith, a physician and his wife. The Griffiths had just bought an antebellum estate called Two Rivers and were beginning to restore the house. According to Captain Long, on the morning of Christmas Eve, Mrs. Griffith was showing a friend the house and grounds when she suddenly noticed something amiss.

In back of the house was a tiny family cemetery where eight members of the Shy family, the mansion's original owners, had been buried in the 1800s and early 1900s. Mrs. Griffith noticed that the most prominently marked grave had been disturbed. The grave's headstone, more than a hundred years old, bore this inscription: *Lt. Col. Wm. Shy, 20th Tenn. Infantry, C.S.A., Born May 24, 1838, Killed At Battle of Nashville, Dec. 16, 1864.*

Beneath the headstone the earth was freshly turned, down to a depth of three or four feet. Grave-robbers, Mrs. Griffith thought, probably searching for Civil War artifacts. She didn't see signs of a coffin, either on the ground or down in the grave itself—perhaps they'd been scared off before reaching it—but she called Sheriff Fleming Williams anyway.

Needless to say, most of Sheriff Williams's deputies were doing what most of the rest of us were doing: enjoying the holidays with their families. The sheriff came out, took a quick look, and—since there didn't seem to be any dire emergency—told her he'd be back after Christmas. A churned-up grave in a tiny old cemetery was nothing to get excited about, he thought.

When he returned on December 29, though, his thinking changed swiftly. Just below the surface of the recently disturbed earth, he found what appeared to be a recent murder victim. More precisely, he found *most* of one: the body had no head.

Sheriff Williams radioed the Williamson County coroner, Clyde Stephens, who hurried out to the Griffiths' backyard, joining what was fast becoming a throng of deputies. Under the coroner's direction, they continued the excavation very carefully, so as not to destroy any evidence they might need in a murder trial.

The body was that of an elegantly dressed young man decked out in a tuxedo of some sort. Although it was pretty ripe, the corpse was still largely intact and its flesh was still pink. The informal consensus was that, whoever he was, he'd been dead no more than a few months. But how had he come to be recently buried, or partially buried, in an old Civil War grave?

Easy, thought the coroner: What better hiding place for a body—a *second* body—than a grave? It was simply a macabre twist on the old trick of hiding something in plain sight. But apparently the killer had been scared away halfway through the task of burying his victim. A grave-tampering incident was one thing; a murder case was quite another. In a hurried graveside conference, the sheriff and the coroner decided they might need some expert help excavating the remains. That's when Detective Captain Long called me.

I told Captain Long that I would meet him at the sheriff's office the following morning and that I would bring an assistant: my son Charlie. While his Arizona classmates were off skiing and partying, Charlie would be getting valuable field experience in a homicide investigation—an enviable Christmas bonus for any aspiring anthropologist.

We set off early, heading west on Interstate 40 in my Mustang convertible. It was a cold, wet day, so, needless to say, we did not put the top down. A few months after I had bought the car, Charlie—who, unlike me, loved speed and was, after all, a teenager at the time—whipped into the left lane of a prairie straightaway just as the farmer he was passing executed a left turn. The Mustang was never quite the same after that.

On this gray December morning, I was at the wheel—not because I didn't trust Charlie's driving but because I tend to get carsick if I'm not steering. During the three-hour drive to Franklin, we talked about Charlie's studies at Arizona. His major professor, Walter Birkby, had been my first graduate student at the University of Kansas, so I got to catch up not only on Charlie's progress but also on Walter's career. The miles passed swiftly.

We arrived in Franklin at about 10:30 A.M. and followed Captain Long out to Two Rivers. After some 125 years the two-story home was obviously in need of its current restoration, but it was still striking: red brick, black shutters, and tall chimneys at each end. Big oaks and maples filled the front yard.

In back, the ground sloped down toward the Harpeth River; on a gentle rise, halfway between the house and the river, a cluster of headstones marked the Shy family cemetery. Directly behind Colonel Shy's stone marker was an oak tree; directly in front was the muddy hole in the ground. As we neared the grave I noticed that the sod had been carefully removed and set aside. I guessed that whoever dug that hole had planned to cover his tracks thoroughly, until something—a barking dog, an unexpected porch light, or possibly even Mrs. Griffith's home-and-garden tour—sent him scurrying away.

The hole measured about three feet square and three to four feet deep. Peering down into it, I could see exposed flesh and bone. With Charlie's help I began to clean out the disturbed soil and expose the body. The ground was wet and the hole was muddy. At first we lay on a piece of plywood positioned at the edge of the grave, reaching in with trowels to pick the earth loose. Except for the cold and the rain, the work was easy, because the dirt had been disturbed so recently. As the hole got deeper I climbed down inside. Over the years, counting my excavations of Indian burials in the Great Plains, I've been in somewhere around five thousand graves. By the time I die I suspect I'll hold some sort of unofficial record: "body that's been in and out of the most graves ever."

Just as Captain Long had told me over the phone, the body was in an advanced state of decay. By now some of the joints had deteriorated. The legs were separated from the pelvis, and the arms were detached from the torso. The knees and the elbows, however, were still intact and still covered with clothing, as was most of the torso. From the look of the formal black jacket and pleated white shirt, I wondered if the victim had been a waiter from some fancy Nashville or Franklin restaurant. Either that or a groomsman at a wedding, a guy who'd indiscreetly dallied with the wrong bridesmaid—or the bride.

The body was in a sitting position on top of the antiquated coffin that had been buried in 1864. From excavating thousands of Native American burials on the Great Plains in the 1950s and '60s, I knew that burying a body in a flexed position required less digging than stretching it out horizontally. It was one more sign of someone hurrying to conceal a crime.

As we dug deeper and exposed more of the body, I saw a small hole in the top of the old coffin. The coffin appeared to be made of cast iron—the top of the line, funereally speaking, back in the 1860s. The hole, which measured about one foot by two feet, might have been caused by the force of a pick or shovel striking the brittle metal. Then, as the disturbed, soggy earth settled around the hastily buried victim, the pelvis and lower spine dropped through the opening and into the old coffin. As a result, I had a hard time extracting the remains.

As I carefully unearthed body parts and pieces of clothing, I handed them up to Charlie, who laid them out in anatomical order on the plywood. Once I'd recovered all the body parts I could find, he placed the pieces in evidence bags and labeled them. In addition to the body, I found two cigarette butts, which Charlie bagged as well.

Over the years I've noticed that killers tend to smoke heavily at crime scenes. In one murder case—involving an auto chop-shop owner who shot a snitch with a hunting rifle—I found a whole pile of mini-cigar butts at the spot where the killer had lain in ambush for hours. Those particular butts had plastic tips, which he'd bitten down on with enough force to leave tooth marks; fortunately, I was able to match those marks with a cast we later made of his teeth. Under the circumstances, chain-smoking is not surprising, I guess—a killer is likely to be very tense, and smoking is a nervous habit—but it's not too smart, either, since even paper cigarette butts can pick up fingerprints and saliva-borne DNA—evidence that can send a killer to death row. (Note to smokers: That's one more way smoking can kill you.)

As I excavated, the hole got deeper and deeper; by the time I had re-covered most of the body, I had reached the top of the Civil War–era coffin. I asked a deputy to loan me his flashlight, instructed Charlie and the deputy to hold my ankles, and hung headfirst in the pit so I could peer inside the hole in the coffin's lid. There wasn't really anything to see—just a thin layer of goo in the bottom—but then again, I hadn't ex-pected there to be anything left after more than a century. Several years before, I had excavated a cemetery dating from this same period, the mid- to late 1800s. That cemetery contained nearly twenty graves, but the bone fragments I recovered from that entire cemetery could fit eas-ily in the palm of one hand: they'd crumbled that completely in the damp dirt of Tennessee. Knowing what I did about Civil War–era buri-als, then, I would have been astonished if Colonel Shy's bones had shown up in the beam of the flashlight. With a grunt and a tug, Charlie and a deputy hauled me up out of the grave.

By now Charlie and I were both soaked and chilled to the bone. We took off our muddy jumpsuits and put them in the trunk of the Mus-tang, along with the remains and the clothing, which we'd removed from the body and bagged separately. Before heading back to Knoxville, we needed to make a brief detour to the state crime laboratory near Nashville, where TBI technicians would pore over the clothing and cig-arette butts for clues to the identities of our victim and his killer.

We got to the crime lab late in the day, just before closing time. The clothing was wet and smelly, so the TBI staff did not welcome us with open arms. To keep from stinking up the entire laboratory, they finally decided to spread the clothing out in their heated garage to dry and air out.

Charlie and I got back to Knoxville late that Friday night. I pulled into the garage—fortunately, it was not attached to the house, so we wouldn't smell the body—and headed inside for a shower, sleep, and a weekend of college football bowl games. Whoever was out there in the Mustang, he wasn't likely to go anywhere, because I took the car keys with me.

On Monday morning I took the remains to the anthropology department offices beneath the football stadium and put them in several large pots of hot water to soften the tissue for easy removal. (By now, after many years and two replacement stoves, I had learned not to do this at home.) The process of sorting, cleaning, and examining the bones would take a couple of days, even though the skeleton wasn't complete.

It wasn't just the skull that was missing; so were the feet and one of the hands. That's common with bodies recovered outdoors: dogs, coyotes, vultures, and raccoons often feed on corpses, and the hands and feet are the easiest parts for predators to pull off and drag away. In this case, though, I wasn't sure what to make of that, since the body had been buried, or at least partially buried. Interestingly, the one hand that remained was still inside a white glove when we found it, reinforcing my sense that the victim might have been a waiter at an upscale restaurant or an usher at a wedding.

I was pretty sure, right from the start, that this was a male; however, the genital region was one of the areas where decomposition had reached the advanced stage, so I knew I'd have to rely on the pelvis and other skeletal indicators to confirm the sex. The pubic bones were short and sharply angled—not the sort of pelvic geometry conducive to childbearing. Clearly, our mystery corpse was a mystery man.

The sternal end of the clavicle, where the collarbone joins the breastbone, was fully fused, so that meant he was probably at least twenty-five. The pubic symphysis—the joint where the pubic bones met at the front of the abdomen—had a rough, bumpy surface, which told me he was probably somewhere in his mid- to late twenties. To check my own conclusions, I called in six of my graduate students—by now, students were filtering back from their holiday travels—and asked them to estimate the man's age. All six put the age at twenty-six to twenty-nine.

The femoral head, the ball at the top of the thighbone, measured 50 millimeters, or about 2 inches, in diameter—also pretty typical for a male. The left femur measured 490 millimeters in length, or about 19.3 inches, and the right femur was 492. Using a formula derived by anthropologist Mildred Trotter and statistician Goldine Gleser in 1958, I calculated that our victim had once stood between five feet nine inches and six feet tall—when he still had his head, that is.

The process of cleaning and examining the bones failed to turn up any indication of the cause of death. As decayed as the soft tissue was in places, we wouldn't have been able to detect stab wounds even if they had been present; the bones themselves bore no cut marks or other signs of skeletal trauma. Judging by the state of decomposition, I still estimated the time since death at a few months, or possibly more, but definitely less than one year.

Police in Williamson County and Nashville checked for missing-person reports filed within the past year. No one at all was missing in Williamson County; none of Nashville's missing persons matched the physical description of these remains: white male, mid-twenties to early thirties, about five feet ten inches tall.

Area newspapers—hurting for juicy news during the lull between Christmas and New Year's—got wind of the mystery and began reporting it. HEADLESS BODY FOUND AT FRANKLIN, read one headline on January 1. The story, sent out over the Associated Press wire service, told how the body was found sitting atop Colonel Shy's coffin. It also described the "tuxedo-type shirt, vest, and coat" and quoted my estimate of time since death: "It appears the man has been dead two months to a year," I said, "and a year may be a little too much." I gave another reporter a narrower range, two to six months.

A day or two later, one enterprising reporter started looking into other recent deaths and found one in Knoxville that bore some similarities: Less than two months before, a decapitated man had turned up in a rural area just outside Knoxville. Could the two cases be related, the work of a serial killer? I told him I didn't think so. The Knoxville victim had been dismembered and mutilated—his head and neck hacked off, his arms and lower legs severed, even his genitals cut off. The Franklin body—at least, what we had of it—showed no cut marks. TORSO CASE NOT LINKED TO OTHER DECAPITATED BODY, the resulting headline proclaimed.

Then, on January 3, the plot thickened: A Williamson County sheriff's deputy arrived, bearing the skull and mandible. The coroner and sheriff's deputies had gone back to the grave, excavated further, and located the skull inside the coffin. "It's my theory that he was crammed head first in the hole made in the colonel's casket," the coroner told a UPI reporter. OFFICER'S GRAVE MYSTERY GROWS, read that day's headline. The story began, "The head, feet and an arm of an unidentified body found in the grave of a Confederate officer have been recovered from inside the officer's coffin, authorities said."

There was no longer any mystery about the cause of death: A gunshot of enormous force had blasted into the forehead about two inches above the left eye; the exit wound—if you could call it that—was at the back of the head, near the base of the skull. I say *skull*, but that's not exactly accurate: The force of the projectile was so great that it shattered the poor man's head into seventeen pieces. I had to glue them back together just to determine the location and size of the entry and exit wounds. Judging by the destruction, he had been shot by a large-caliber gun, possibly at close range. Our mystery man had died a violent, instantaneous death.

The latest wrinkle in the case was this: Unlike the rest of the body, the skull was virtually fleshless and chocolate-brown in color, much like the ancient Indian skulls I'd excavated in South Dakota. The teeth had no fillings but lots of cavities, some of them quite large; his lower-left third molar was on the verge of abscessing. There was no indication that this elegantly attired gentleman had ever set foot in a dentist's office or had ever received a scrap of dental care—modern dental care, anyway.

An uncomfortable suspicion began gnawing at me.

Just then the telephone rang. It was a technician from the state crime laboratory in Nashville calling. "Dr. Bass, we're finding some odd things in this clothing you brought us," he said. "The fibers are all natural—cotton and silk; nothing synthetic." There were no labels in the clothes that could be traced, he added, and the trouser legs, which laced up the sides, were unlike anything he'd seen before. The square-toed shoes were a style that had become popular a few years earlier—but were also a style that had been common a century before.

His final question was the one I'd suddenly guessed, with a rush of dread, might be coming: "Do you think this could actually be the body of Colonel Shy?"

"I'm starting to think that it is," I admitted. I was glad he couldn't see my face turning crimson with embarrassment. "I still have a few questions I need answers to—for example, did they have elastic like what's in those shoes back in 1864?—but it's looking more and more likely."

There's a time-honored philosopher's maxim—Occam's razor, it's called—that holds that the simplest explanation that fits the facts is usually right. Over the years I'd seen enough bizarre twists in murder cases to know that Occam's razor could sometimes cut the wrong way, but in this case it seemed right. If the body in my lab was that of Colonel William Shy, it would answer a lot of questions: Why were the cavities in the teeth unfilled? Why did the clothing look not just so formal but so unusual? Why were there no synthetic fibers, no labels, no other traceable artifacts?

When we found the body sitting atop the coffin, it looked as if it had been *added* to the grave, not pulled out of a small hole in the coffin's lid. Having assumed it was an additional body, we easily took the next logical step: it must be a murder victim, and a recent one at that. Our next feat of deductive gymnastics—explaining away the absence of a body within the coffin—had been easy, in light of my prior excavation of tiny fragments from a nineteenth-century cemetery. (Clyde Stephens, the coroner, explained the absence of a body another way, voicing doubts that Colonel Shy had ever occupied the coffin in the first place: "I would have thought there would have been possibly a belt buckle, buttons, or something," he told a Nashville reporter, "but we didn't find anything.")

At least, we didn't find anything where we'd expected to. To the embarrassment of everyone involved—or at least everyone quoted in the press—it now appeared that it was Colonel Shy himself who had been hiding in plain sight. Instead of a recent murder victim crammed partway into a coffin, the body was an old soldier pulled mostly *out* of the coffin, losing his head and some appendages in the grave-robbing tug-of-war. The shattered skull made perfect sense in this new light too: Colonel Shy was killed when Union troops surrounded and overran the hilltop where the 20th Tennessee Infantry had sought safety. The colonel fell in fierce hand-to-hand combat, shot with a .58-caliber minié ball in the forehead at point-blank range.

By now the story had mushroomed from a local crime story into a human-interest feature on the worldwide Associated Press wire service: A mysterious corpse baffles police; they consult a prominent scientist; the scientist errs spectacularly; the ancient soldier has the last laugh. Judging by the letters and phone calls I got, the story was picked up by papers everywhere. One former student sent me a copy from an English-language paper in Bangkok, Thailand.

A few weeks later Colonel Shy was reburied in his grave. A local funeral home donated a new coffin, and a regiment of more than a hundred Civil War reenactors turned out in full uniform to give him a full military burial. As the minister concluded his graveside remarks, lightning flashed, thunder rolled, and hailstones pelted the crowd—exactly as historical accounts said they had at the colonel's first funeral, 113 years before! This time, perhaps, the Confederate soldier could rest in peace.

I, on the other hand, could not. Although identifying the body as Colonel Shy's had answered several questions, it had raised one enormous new one: *How could I have misjudged the time since death by the whopping margin of almost 113 years?*

That question, it turned out, had several answers. The first and simplest answer came to light when we subjected a tissue sample to chemical analysis. The body, it turned out, had been embalmed—not nearly so common in the 1860s as it is today, but not too surprising for an officer and a gentleman of wealth and social prominence. A man of Shy's standing would have been buried in his best clothes—the very same black jacket and pleated shirt that we later recognized in the last known photo of Colonel Shy, taken in the early 1860s.

The next piece of the puzzle took some metallurgical and chemical detective work. The coffin was cast iron, remember, so stout that it kept out water for more than a century. It also kept out the coffin flies—tenacious, gnat-size flies that can burrow deep into the ground and bore through wooden coffins and penetrate tiny openings in metal coffins. And because the coffin was hermetically sealed, there was very little oxygen for bacteria to draw on to digest the body's soft tissues—hence, the pink tissue that appeared to be only two to six months postmortem.

Those were partial answers to the troubling question I'd asked myself. The more comprehensive answer was also more unsettling: I just didn't know enough—not nearly enough—about the postmortem processes that begin when human life ends. And it wasn't just me: *None of us* knew enough. Anthropologists, pathologists, coroners, police—we were all woefully ignorant about what happens to bodies after death, and how, and when.

Colonel Shy—ably assisted by a few newspaper reporters and my own big mouth—had revealed both the depths of my own ignorance and the huge gap in forensic knowledge. Personally, I was embarrassed; scientifically, I was intrigued; above all, I was determined to do something about it.

From that moment on, everything would change, in ways I could never have imagined.

CHAPTER 6

The Scene of the Crime

FOR REASONS I don't fully understand, forensics has suddenly become a hot topic on television. Night after night, a seemingly endless parade of victims is murdered, and night after night those murders are swiftly and cleverly solved. On most television dramas, at least, the forensic scientist is practically a god, endowed with a huge intellect and outfitted with every razzle-dazzle technology imaginable.

It pains me to admit it, but I am somewhat less brilliant than TV supersleuths—and, with all due respect, so are many of my forensic colleagues. We're not geniuses, and our gadgets can't answer every question or pinpoint every perpetrator. But even though TV sometimes creates unrealistic expectations about the swiftness and certainty of murder investigations, some shows have done a great service by spotlighting the role forensic scientists—even ordinary, real-life ones—can play in bringing killers to justice. And these shows do get a lot dead right: Crime scene investigation is absolutely crucial to solving a crime.

Surprisingly, many of my fellow forensic anthropologists—probably nine out of ten—have never worked a crime scene. They're happy to examine bones on a lab table or under a microscope, but they don't dirty their hands or shoes in the muck, mud, or blood of fieldwork. They stay clean and dry that way, but they also miss a lot of evidence that could reveal the truth about what happened to a murder victim. A victim like James Grizzle, whose story—as we pieced it together at the crime scene—is one of the most bizarre and shocking I have ever encountered.

One chilly January morning, I got a phone call from a detective with the Hawkins County, Tennessee, sheriff's office, asking if I could help search for the body of a man whom they suspected had burned to death in his house a week or so earlier. I agreed to help, and enlisted three of my brightest graduate students—Steve Symes, Pat Willey, and David Hunt—to make the hundred-mile trip to Hawkins County the next morning.

By now I'd been searching crime scenes and death scenes in Tennessee for ten years, and I'd developed an approach that seemed to work quite well. Anytime I received a request from law enforcement for help finding, recovering, or identifying human remains, I took a four-person forensic response team: a faculty member (me in those days, though now other faculty members take turns taking forensic cases) and three students trained in osteology, identification of human bones.

I no longer used my own car. The anthropology department now had a pickup truck, which we kept loaded at all times with the equipment we'd need in the field—shovels and trowels for digging; wire-mesh screens for sifting small bones and bone fragments from dirt; three body bags for transporting corpses in the back of the truck (beneath a camper shell); paper evidence bags for collecting scattered bones, bullet casings, cigarette butts, beer bottles, knives, and any other evidence we recovered; one-hundred-foot surveyor's tapes for measuring the proximity of bodies or bones to fixed landmarks such as trees, utility poles, and buildings; either red or orange survey flags for marking the location of every bone or piece of evidence; and at least two cameras.

I considered the cameras the most important part of our equipment; they were essential in documenting the scene, the search, and particularly the recovery of human remains. I know of only two types of scientific research that require utterly destroying the very thing you're studying: excavating an archaeological site and investigating a death scene. By the time you're finished, it's gone, dismantled, so you better make damned sure you've got an exhaustive record on film; there's no going back to check for something you overlooked—say, footprints on the surface of a shallow grave—after you've trampled or dug up the ground.

It was Kansas lawman Harold Nye—a living legend at the KBI—who taught me one of my most important lessons about crime scene investigation: "Shoot your way in, and shoot your way out." It sounds like the modus operandi of a trigger-happy bank robber, but Harold was talking about photography. "When you arrive at the scene and get out of your car, take a picture of the house or the car or whatever the scene is," he said. "As you walk closer, take some more. Take pictures of the ground before you walk on it; take pictures of who's there; pictures of what kind of shoes officers at the scene are wearing. Take pictures of the body before you move it or even touch it."

Harold had shot his way into the Clutter family's farmhouse the night the bodies were discovered there. If he hadn't—if he or anyone else involved in the investigation had set foot in the basement before Harold photographed its dusty floor—the KBI would never had seen and preserved on film the footprints that were later linked to the killers' boots. Because Harold shot his way in, the telltale footprints were caught on film and the killers were convicted.

It's hard to put a price tag on human life and criminal justice; film, on the other hand, is pretty damn cheap. Over the decades I've taken hundreds of thousands of crime scene photos, and I've never regretted a single click of the shutter. As cameras become more and more sophisticated—exposing for infrared frequencies or heat, capturing high-resolution digital images, and even incorporating GPS (global positioning system) receivers that automatically record precise location coordinates in longitude and latitude—photography will sharpen the focus of crime scene investigation still further.

On my four-person forensic teams, one member always served as our photographer. For the search of the burned house in Hawkins County, the camera would be wielded by Steve Symes, one of my Ph.D. students. Steve had shown a remarkable talent for crime scene photography; his photos often revealed far more detail than those taken by the official photographers from police departments or sheriff's offices. On this day, although I didn't know it at the time, Steve would be laboring under a severe handicap: That morning he'd awakened severely hungover, chilled to the bone, and wringing wet. Sometime in the night, after an intoxicated Steve had fallen asleep, his water bed had sprung a leak, dumping dozens of gallons of water across his floor and through his downstairs neighbor's ceiling. Luckily the wiring in his electric blanket was waterproof; otherwise he might have been fried. As it was, he felt sick as a dog, and the mountainous roads of eastern Tennessee weren't helping any.

It took us about ninety minutes to make the drive from Knoxville to the Hawkins County sheriff's office in Rogersville; from there we followed a deputy—a Lieutenant Alvis Wilmot, who was heading the investigation—out a winding road along the north fork of the Holston River.

When you're out in the country from Rogersville, population four thousand, you're pretty far out there and pretty close to nowhere. By the time we turned down a gravel drive about twenty-five miles outside town, we were in a remote river valley, so sparsely populated—or maybe so suspicious of outsiders—that the fire hadn't even been reported until a relative of the house's owner drove down from Virginia and found the place in ruins. The property was heavily wooded and steeply sloped, angling down on the east side to the clear, green waters of the Holston's north fork. We all got out and stretched our legs; Steve took some particularly deep breaths.

According to Lieutenant Wilmot, the blaze had occurred eight days before; as best they could tell from their interviews with the nearest neighbors, it probably began around two o'clock in the morning. By the time it had burned itself out, all that remained was a rectangle of charred rubble, bordered by a jumble of blackened bricks; a larger pile of bricks marked a spot near the center where a chimney had stood.

The house and land had been purchased just a month or so earlier by a Virginia man named James Grizzle, who hailed from an area even more mountainous and less populated than this one. Grizzle had moved into the house in December to begin remodeling it. The fire occurred on January 15; six days later, not having seen or heard from his son, Grizzle's father had come looking and had promptly called the sheriff upon seeing that the house had burned. Our goal was to determine whether Grizzle's body lay somewhere in the charred ruins left by the fire.

FORENSICALLY, fire scenes pose an interesting combination of circumstances and challenges. As at any death scene involving a decomposed body or bone, it's important to locate and recover all the human remains; at a fire scene, though, that's often difficult because of the dramatic changes a human body undergoes in an intense fire.

The arms and legs are the first to go. Relatively thin and surrounded by oxygen, they're like kindling, easy to ignite and quick to burn. At temperatures of only a few hundred degrees, the skin quickly blackens, the fat beneath the skin starts to sizzle, and within a matter of minutes the skin splits open and the flesh begins to burn. As it does, something remarkable and eerie happens. The limbs begin to *move*—the hands and feet clench, the arms curl up toward the shoulders, and the legs spread slightly apart with the knees flexed. It's a function of biomechanics and muscle strength: The flexors, the muscles that cause our arms and legs to bend, are stronger than the extensors, the ones that cause our limbs to straighten. As fire cooks and dries out the muscles and tendons of the body, they shrink, just like a steak on the grill, and the flexors overpower the extensors.

The resulting position is very much like a boxer's stance in the ring; for that reason we call it the "pugilistic posture." It's very distinct and very consistent—as consistent in fire victims as a purplish color and swollen tongue are in hanging victims—so long as the limbs are free to flex. If, on the other hand, the arms are tied or pinned behind the back, they won't be able to curl up, so finding a burned body whose arms are straight can be an important clue that the victim was somehow confined or restrained.

The other truly dramatic change that occurs is to the head. The skull is basically a sealed vessel, filled with fluid and moist brain tissue. It doesn't take long for all that moisture to reach the boiling point and create pressure in the cranium; the hotter the fire, the greater the pressure. If there's an outlet for that pressure—for example, a bullet hole in the skull—the pressure vents harmlessly. If there isn't, the skull can literally burst, fracturing the cranium into numerous pieces, each about the size of a quarter. Recovering and reconstructing a skull from a fire scene is one of the most tedious tasks a forensic anthropologist ever faces, and even after it's pieced together, that skull remains a challenge, since blunt-force or sharp-force trauma can be difficult to spot amid the myriad fire-induced fracture lines and the occasional gaps where pieces are missing.

Fortunately for crime scene investigators, it's difficult to burn up a body entirely; even cremation leaves substantial portions of bone, which must then be pulverized mechanically. Still, even the biggest, most robust bones of the body—the femur and tibia in the leg, the humerus in the arm—can be badly damaged by a fire. A fairly low-temperature house fire will turn the long bones black or caramel-colored but leave them relatively intact structurally. An arson fire, though—one fueled by gasoline or some other flammable accelerant—can reach temperatures as high as 2,000 degrees Fahrenheit; at such extreme temperatures the bone undergoes a chemical and structural metamorphosis. Bone, like the rest of the body, contains carbon, and at extremely high temperatures that carbon burns out of the bone. What's left behind, called "calcined" bone, might still retain its shape—just as a coral reef retains its form even after the organisms that built it die—but it will be very lightweight, grayish in color, riddled with heat fractures, and so fragile that it can crumble in your hands, and will certainly crumble underfoot. (Recently, I was contacted by an attorney who's preparing for a retrial of a murder case; he told me that a key piece of prosecution evidence—a calcined piece of the victim's burned skull—was accidentally dropped on the floor and stepped on by a judge, reducing it to powder.)

For all their destructive power, fires leave behind a surprising amount of evidence, though you have to know where and how to look for it. Actually, I've come to enjoy the challenge, the scientific puzzle, of mentally reconstructing what a fire scene looked like just before it burned. Those buttons and snaps, hooks and eyes, brass rivets and zippers, embedded in a heap of ashes? Easy: a chest of drawers once crammed with shirts and brassieres and blue jeans. That mound of broken glass and porcelain beside a charred chandelier? It was once the china cabinet in the dining room.

The key to mentally reconstructing a burned house is the careful sifting through a layer of ash several inches thick—the remnants of the ceiling and roof. Beneath that layer is a wealth of information about the way things were. For instance, most chairs in houses are made of wood, but they usually have small metal feet on each of their legs, which can indicate their position at the time of the fire. A desk might burn, but the paper clips and staples will mark its location; a cache of needles, pins, and scissors might have once belonged inside a sewing basket.

The most valuable thing I have found at a fire scene was a $12,000 diamond necklace. It was a woman's Christmas gift from her husband, unwrapped just months before she burned to death in a suspicious fire in their mansion. When I found the necklace—at the base of a wall, beneath a layer of ash—it had a safety pin fastened around it. That puzzled me, and so did the location where I found it, so I asked her family if they could shed any light on either question. Her relatives told me that she liked to pin her jewelry into the folds of her drapes; when the drapes were closed, the jewelry was on display; when they were open, the jewelry was hidden. Sure enough, I'd found it directly under a window. The explanation matched what we found at the scene.

Sometimes what you *don't* find at a fire scene tells you as much as what you *do* find. I once excavated a fire scene that had already been examined by the police and an arson investigator, none of whom noted anything suspicious. What struck me most about the house, as I recovered the incinerated body, was that there were no dishes or silverware in the kitchen, no coat hangers in the closets, no picture frames or hangers on the wall. (Pictures themselves will burn, and so will wooden frames, but metal frames, and even the little screws and nails and wires on the back of a wooden frame, don't burn; they fall to the floor at the base of the wall.) To me it was obvious that the house had been stripped bare, except for a few large items, before the fire—a classic indicator of arson. But the strangest part of the story, as well as we could reconstruct it, was this: The dead man wasn't the homeowner but the man hired to burn down the house; apparently, as he was dousing the structure with gasoline—during a severe thunderstorm, we learned—lightning struck the house, igniting the gasoline vapors in a fiery explosion that killed him almost instantly. It was one of the best cases of bad timing I've ever seen. In this case the evidence at the scene revealed that crimes had indeed been committed, but they were arson and insurance fraud, not murder.

Anytime I'm called to a fire scene, I try to find all the skeletal material, but I don't stop there; I also deduce as fully as possible the events that happened before and during the fire. I pay particular attention to identifying jewelry, teeth, and bones, but I also check and recheck for other evidence, and I consider all that evidence before I draw any conclusions about what happened.

The single thing that does the most to destroy forensic evidence at a fire scene is not the fire itself; it is an untrained, overzealous investigator armed with a rake. An investigator who hasn't been trained in human osteology and doesn't know how to recognize and identify burned bone fragments can wreak havoc with a fire scene. It's maddeningly common for police who are looking for a body to go over an entire scene, raking all the burned material into long ridges, or windrows, about three feet apart. Think about it: If you want to know the location and arrangement of a body when the fire began—and if you want to know its proximity and placement relative to items such as a gun, a knife, or bullets—what hope do you have if you stir everything up with a rake?

I once arrived at a fire scene with a team to search for the body of a suspected suicide victim, only to be told by a fire marshal that I needn't bother looking. The scene was massive—a farm compound consisting of a house, a barn, and half a dozen other outbuildings; the firefighters and arson investigator had used a backhoe to clear out portions of the rubble. I figured the most promising place to search was the house, but the fire marshal scoffed at me. "We've raked that house five times," he said. When I allowed as how we'd like to take a look anyway, seeing as how we were already there, he shook his head and walked away as if we were idiots.

Sifting through the churned-up mess, we found a few pieces of a man's skull. There was barely a handful of fragments left—when you roll a backhoe over calcined bones and then flail away at them with a bunch of rakes five times, you're going to pulverize things pretty thoroughly— but it was enough to indicate that the man had set fire to his homestead and killed himself.

FORTUNATELY, in the Hawkins County case, the sheriff's office had called us before the fire scene had been disturbed; the arson investigator would be joining us there, but we'd get first crack at the scene. If there were burned bones somewhere amid the rubble, we should be able to find them, and they'd probably still be very close together.

On the east, or downhill, side, facing the river, the house had been two stories high; the west side was notched into the hill, with only the main floor above grade. According to Lieutenant Wilmot, the bedroom where Grizzle was most likely to have been sleeping—going by descriptions from the prior owners—was at the north end of the upper floor. Now, of course, there *was* no upper floor: during the fire, the floor joists had burned through and the main floor and roof had collapsed onto the concrete slab running beneath the entire structure. That concrete slab, by the way, was our friend. A smooth, solid surface ringed by low ridges of jumbled brick, it was now one giant evidence pad, saving everything for us.

We started at the downhill face of the house at about 10:30, sifting and probing our way delicately toward the center of the house. At about 11:15, Steve Symes's keen photographer's eye, bleary and bloodshot though it was, zoomed in on a bone jutting from beneath a pile of bricks, a collapsed section of the chimney. As we lifted off the bricks, we found both sets of leg bones and most of the spine. Some of the joints were still partially articulated, or held together by ligaments and cartilage, but many of the bones themselves had been reduced to fragments. Completely calcined, these shards of a shattered life clinked together in my hand like bits of a smashed ceramic mug. This corpse had been seriously incinerated.

The condition of the bones indicated a hot fire. The condition of the electrical wiring confirmed it: the copper had melted, dribbling into ragged lines on the concrete floor. Copper's melting point is around 2,000 degrees Fahrenheit, so the fire raged hotter than that. What's more, such high temperatures clearly pointed to the presence of accelerants; without the addition of gasoline or some other flammable liquid, experiments have shown, house fires usually don't exceed 1,600 degrees.

The concentration of bones was about a foot inside the east wall of the house—the side facing the river—and was several feet north of a concrete-block wall that divided the house into a north end and a south end. As we were removing the bones, we found a mass of burned tissue resting on a piece of white cotton fabric from a man's jockey shorts and a charred pair of olive-drab pants.

At this point we were pretty sure we'd found the body of a male, very likely the missing James Grizzle. But as we continued to search the scene, the picture got fuzzier, not clearer, and more and more intriguing.

The position of the legs, pelvis, and spinal column indicated that the body was lying on its back; the legs were bent or folded over the top of the body, with the knees up above the shoulders—occupying the space where the head should have been but wasn't. We explored the surrounding area thoroughly in search of the head. Finally, about six feet away, embedded in another pile of bricks, we found arm bones, a few ribs, and the skull and mandible. These bones, like the first batch, were oddly arranged and badly fragmented, apparently from the fire.

But why were they six feet away from the lower two-thirds of the body? As I sorted through the possibilities in my mind, I considered the fact that the house was a two-story structure. I have seen cases, in similar buildings, where part of a body has burned and fallen through a hole in the floor, leaving the other part to settle elsewhere, atop a different layer of rubble. Could that have happened in this case?

I looked again at the legs and pelvis. Besides the fabric of the underwear and pants, there wasn't much under the bones—just some unburned Sheetrock or drywall, unburned floor tile, and the house's concrete slab. There was also very little beneath the head, arms, and ribs. If one part of the body had burned away and fallen through a hole in the upper floor, leaving the rest of the body up in the master bedroom until the entire floor collapsed, we should have found quite a bit of burned debris underneath one of our groupings of bones: the charred remnants of wooden joists, subflooring, and flooring material—maybe even blackened bedsprings and a burned mattress, if the man had been asleep at two A.M., when the fire began. The fact that so little other material was beneath the bones suggested that the entire body was probably already down in the basement when the main floor burned through and collapsed onto the slab.

But if that was the case, why on earth was the top of the body so far away from the lower portion? I've seen many cases where the intense heat of a fire caused a skull to burst or shatter, but I've never seen one where it caused the head and upper torso to fly across a room.

As I stood there scratching my head, looking from one heap of bones to the other, I said—thinking out loud, mainly—"The only thing I can think of that would explain this separation is some sort of explosion."

As soon as I said it, Lieutenant Wilmot spoke up. "Funny you should say that. One of the neighbors down the road said he heard an explosion before the fire." It would have saved me some puzzlement if he'd thought to pass along that investigative tidbit a little sooner; on the other hand, if he had, I wouldn't have had the fun of formulating an exotic theory. I inspected the bones again. The surface of the sternum was badly fractured and pitted; the spine had separated just below the skull— exactly where it would if a violent explosion had ripped apart the chest.

The fragmentation of the body was not the only indication of violence. Several inches from the spinal column, in the region of the thoracic vertebrae and ribs, we found an oblong disk of lead. Measuring about an inch long and three-quarters of an inch wide, it was flat on top. Its underside bore the impression of woven fabric. It didn't take a forensic genius to guess that before the fire, and before the explosion, there had been a gunshot. From no more than a few feet away, a bullet had been aimed at a human heart.

Some things remained puzzling, but one thing was clear: unless the victim had carefully doused the house with gasoline, strapped a stick of dynamite to his chest, lit the fuse, and then fired a gun at his heart, this was a clear-cut case of murder by a killer who had gone to great lengths to destroy the evidence of his crime. Great lengths, but not successful ones.

Working steadily as a team—David Hunt and I excavating material, Pat Willey diagramming our finds and bagging the bones, Steve Symes taking photo after photo—we plucked and sifted bones and teeth from the ashes. As the cold afternoon winter light began to fade, we loaded up for the two-hour drive back to Knoxville. Some twenty paper bags of burned remains lay in the back of the truck, and two tantalizing questions hung in the air between us: Were these the bones of James Grizzle? If so, who had killed him, and why?

Answering the first of those questions required a close examination of the bones and teeth. At the scene, we'd been pretty sure the remains were male. The long bones were quite large and robust, and although the skull was fragmented, the external occipital protuberance—the bump at the base of the skull—was easily identifiable and unusually massive, an almost certain indication of maleness. Measurements in the lab corroborated this further: The head of the femur—the ball that inserts into the hip socket—usually measures 45 millimeters or more in diameter in adult males; our victim's femoral heads measured a whopping 50 millimeters, or almost 2 inches. The circumference of the femoral shafts was also quite manly, at 94 millimeters; a woman's femur rarely exceeds 81 millimeters in circumference.

To determine race, we looked at the facial structure. Although the skull was badly fragmented, portions of the upper and lower jaws were intact enough to interpret. The alveolar areas of the mandible and maxilla, where the tooth sockets met the jaws, were flat, and the teeth were perpendicular to the jaws, rather than jutting forward. The jaws, in other words, were a white man's.

Our victim was clearly an adult. His collarbones, or clavicles, had fully fused or matured, so we knew he was at least twenty-five years old. His lower spine showed the beginnings of osteoarthritic lipping—ragged, jagged shelves projecting from the edges of vertebrae—suggesting that he was over thirty; however, the lipping was slight enough to indicate that he was probably not more than forty. Lieutenant Wilmot had told us that James Grizzle was thirty-six years old, so at this point the smart scientific money was betting that this was indeed Grizzle. To be sure, though, we'd have to get lucky with dental records.

Before moving to the Bible Belt of Tennessee, Grizzle had been a steelworker in the Rust Belt of Indiana. An employee of Bethlehem Steel, he'd had good medical and dental benefits—and a conscientious dentist in La Porte, Indiana, who had taken X rays a few years before.

The mandible, or lower jaw, is denser bone than the maxilla, or upper jaw, so it had emerged from the ashes more intact. In both jaws, though, the heat of the fire had shattered most of the teeth at the junction where the enamel meets the root. By and large, then, we couldn't look for fillings; we'd have to match distinctive features in the structure and geometry of the tooth roots and the jaws themselves.

Grizzle's mandibular X ray showed us the following: His left third molar—his wisdom tooth—was not fully erupted; his left first molar was missing, with bone beginning to resorb or fill the empty socket; his right first molar and right second molar sockets were also empty and beginning to fill in with bone. (His dental-care benefits might have been excellent, but his lifelong dental hygiene, or at least his overall dental health, was quite poor.)

Grizzle's maxillary X ray revealed that the left first premolar bore an odd root, in the shape of an S; that same tooth also had a filling on its inner surface.

Fortunately for us, our victim's upper-left first premolar was one of the few teeth whose crown had *not* shattered; on that crown was a filling, exactly where the X ray told us to look for one. The other features— the missing molars, the resorbed bone, and the S-shaped root—all matched perfectly. I called Lieutenant Wilmot to tell him we had positively identified the victim as James Grizzle.

The remaining questions—who had killed Grizzle, and why?— fell to Lieutenant Wilmot and his colleagues to answer. It didn't take them long.

One of Grizzle's neighbors—those concerned, caring neighbors who hadn't bothered to report the explosion or fire at the time—told deputies that after Grizzle bought the house, he had hired someone to help him remodel it. The worker, a man named Stephen Leon Williams, had moved into the house with Grizzle, bringing along his girlfriend for company.

Grizzle had a lot of money in the bank, his father told police—about $30,000 in his checking account and another $9,000 in savings; apparently he made the mistake of telling Williams about it, for the prosecutor alleged that Williams forged Grizzle's signature on checks drawn on the account in the days after Grizzle's disappearance.

As if the murder weren't already gothic enough, one night not long after Grizzle's mangled body was uncovered, a bizarre new twist came to light: An acquaintance of Williams's named Anthony Layne Flynn sat drinking in a Kingsport tavern called Ralph's Bar. His tongue loosened and his judgment impaired by one too many beers, Flynn told his astonished bar mates how Williams had enlisted his help by asking that he bring his Doberman to Grizzle's house to eat the body. But either the dog wasn't hungry enough or the body wasn't yet ripe enough, because he turned up his pointed nose at it.

That's when Williams resorted to dynamite. But instead of decimating the body, the explosion only ripped it in two. Finally, as a last resort, he doused the house with gasoline and set it afire. As the flames roared into the night sky, he must have thought they were covering his tracks completely, destroying all evidence of the carnage he had committed. In fact, the fire was calling attention to it. It was a beacon, blazing brightly in the dark woods, and its message was clear: *Crime scene—investigate with care.*

IN OCTOBER OF 1981, Stephen Leon Williams was found guilty of first-degree murder in the death of James Grizzle. His codefendant, Anthony Layne Flynn, who owned the finicky Doberman, was acquitted and released.

Because of the shocking ways he had desecrated Grizzle's corpse, Williams was sentenced to die in the electric chair. His execution was scheduled for April 16, 1982. His lawyers promptly appealed the death sentence. A series of appeals, then a nationwide moratorium on executions, delayed the sentence year after year.

In 1999, from behind bars, Williams filed a lawsuit against me. His suit named several codefendants: the investigators, a TV production company, and the Discovery Channel, which had featured the Grizzle case in a forensic documentary. I found it astonishing that our legal system would even permit such a thing: Long after his trial, a convicted killer actually sues the people who uncovered and reported the murder he committed. Fortunately, Williams himself voluntarily dropped me from his lawsuit.

More than twenty years after his conviction for murdering, dismembering, blasting, and burning James Grizzle, Williams remains alive and well in a Tennessee prison. As for the crime scene, it has long since been reclaimed by the Tennessee woods. Somewhere on a steep hillside above a ribbon of green water, a deepening layer of leaf litter and silt nurtures a growing colony of weeds, vines, and tree seedlings. Beneath it all, slowly disappearing from view, is a slab of stained concrete and a jumble of bricks. Here, real-life crime scene investigators once sifted down through ashes and came up with the truth.

CHAPTER 7

Death's Acre:
The Body Farm Is Born

> If the victim has already been dead for a long time, the
> head and face will be swollen, the skin and hair will have
> come off, the lips and mouth will be opened, the eyes
> will be protruding, and maggots will be feeding.
>
> —Sung Tz'u, *The Washing Away of Wrongs*,
> Chinese forensic text published in A.D. 1247*

WHEN I REALIZED I had misjudged Colonel
William Shy's time since death—by 112 years,
no less—my first reaction was profound embarrassment. I had made such confident pronouncements
to the newspaper reporters who were following the story,
and I had a lot of words to eat afterward—words that had
been printed everywhere from Tennessee to Thailand.

Humbling experiences can open the door to life's
greatest insights, though, if we're willing to learn from
them. It didn't take long for my personal embarrassment
to give way to professional curiosity. One reason forensic
cases have always appealed to me is the challenge they
pose: They're often tragic crimes, but they're also scientific puzzles to be solved. I've never liked hunting—the

* Translated by Brian McKnight; © University of Michigan Center for Chinese Studies, 1981. Quoted by kind permission.

idea of killing animals for sport has absolutely no appeal to me—but the excitement of unraveling a forensic riddle is probably not so different from the thrill a big-game hunter experiences while stalking a deadly predator.

But just what was the riddle here—what would I be chasing in this case? The more I thought about it, the more exciting it became: my prey would be death itself. To understand fully what had happened to Colonel Shy—and what eventually happens to us all—I would need to track death deep into its own territory, observe its feeding habits, chart its movements and timetables.

More than seven hundred years ago, a Chinese official named Sung Tz'u compiled a remarkable handbook for forensic investigators. The book, whose title is often translated as *The Washing Away of Wrongs*, suggests an impressive array of postmortem examinations and tests that should be conducted during the early postmortem interval—the hours or days following a suspicious death. The book also describes, in graphic terms, the changes bodies undergo during the extended postmortem interval—the weeks and months it takes for a corpse to transform from flesh to bare bone.

In the three-quarters of a millennium since Tz'u's writing, however, virtually nothing more had been discovered or published about the extended postmortem interval. When I examined Colonel Shy's remains in 1977, I had no more knowledge or scientific literature to draw on than Sung Tz'u had possessed in 1247.

Already—long before I made Colonel Shy's acquaintance—the idea of making a scientific study of decomposition had been germinating in some recess of my mind. The seed had been planted back in 1964, when I wrote to Harold Nye at the Kansas Bureau of Investigation and suggested that we find some rancher to help me study decomp on the hoof ("If you have some interested farmer who would be willing to kill a cow and let it lie . . ."). That seed was still lying dormant in 1971, when I moved to Knoxville to head the anthropology department at the University of Tennessee. Along with my new teaching position, the move to

UT brought me a state-level political appointment as well: I was named Tennessee's first (and so far its only) state forensic anthropologist. Even as I labored to sort and stack hundreds of boxes of Arikara Indian bones in the musty offices beneath Neyland Stadium, the letter of appointment arrived. It was a testament to the importance of networking.

A year or two before, one of my University of Kansas Ph.D. students, Bob Gilbert, had requested pubic bones from medical examiners around the country. Bob was researching skeletal differences between males and females—specifically, the gradual changes that occur in the female pubic symphysis, the joint where the two pubic bones, arching forward from the hipbones, meet at the front of the pelvis. In young adults the surface of the pubic symphysis is rugged, ridged, and grooved; by the mid-thirties the bone is denser and its texture smoother; after age fifty, the face of the joint itself begins to erode. Bob's Ph.D. dissertation aimed to chart those changes in the female pubic symphysis in detail, so that anthropologists could estimate age more precisely. To do that, he needed pubic bones, and lots of them.

Some of the medical examiners he contacted were shocked by his request and refused. But Dr. Jerry Francisco, the chief medical examiner in Tennessee, was intrigued by the research and recognized its potential contribution to forensic science. He sent Bob a batch of pubic bones, and he became a good friend of mine, trading stories with me at forensic meetings.

When I told Jerry I was moving to Tennessee, he asked if I'd be interested in joining his staff as state forensic anthropologist. It wouldn't pay much—a flat fee of $150 per case—but the work promised to be fascinating. Immensely flattered, I said yes at once. Not long after that, I also received a fancy badge as a special consultant to the Tennessee Bureau of Investigation. Eventually, I realized that if I hadn't been working these cases as a state official, I could have charged a hefty hourly consulting rate. Sadly, by the time I figured that out, I'd grown far too fond of the fancy title and the shiny badge to give them up just for something as common as money. One particularly complex forensic case in the

1990s consumed hundreds of hours of my time; in that case my $150 fee translated into less than a dollar an hour. But hey, I also got the privilege of taking a lot of abuse on the witness stand too. Defense attorneys love to bring up the Colonel Shy case, even if it has no connection with their client's case, to sow seeds of doubt in the minds of jurors. ("Isn't it true, Dr. Bass, that your estimate of time since death in that case was off by nearly 113 years?!")

I was still settling into my first semester at the University of Tennessee when the calls and cases and bodies began coming in. It didn't take long to notice a difference between Kansas bodies and Tennessee bodies. More often than not, the Kansas corpses tended to be clean, sun-bleached skeletons, like something you'd see in a Hollywood western. The typical Tennessee body, I noticed very quickly, was more often a rotting, maggot-laden mess. In fact, of the first ten sets of remains brought to me for examination by Tennessee law enforcement officers after I arrived in Knoxville, half were swarming with maggots.

The difference was a function of geography and demographics: Kansas is twice the size of Tennessee—about 82,000 square miles, compared to Tennessee's approximately 42,000—but has barely half as many people. Statistically speaking, then, the odds of stumbling across a fresh body in Kansas are only one-fourth the odds of tripping over a corpse in the Volunteer State. (Actually, the difference is even larger than that, because Tennesseans tend to die younger, thanks to a homicide rate that's twice as high—a problem for someone in another field to figure out.) Since there are far more bodies in Tennessee lying around waiting to be found—often by hunters tramping around in the woods—it stands to reason they're likely to be found more quickly than that handful of Kansas bodies quietly skeletonizing out there on the vast, lonesome prairie. Hence, dead Tennesseans are likely to be a lot messier and smellier.

Still, there was justice to be served. And for a forensic anthropologist—particularly a TBI-badge-toting, official state forensic anthropologist—squeamishness was not an option. I had let it be known that

I was available to help identify bodies or determine cause of death. Therefore every case was welcome and so was every body. But some were more welcome than others—to me, and to the other faculty and staff who shared our quarters beneath the football stadium. It was the janitor who finally snapped.

A FISHERMAN HAD found a "floater"—a floating corpse—in the Emory River, about fifty miles from Knoxville, and a Roane County deputy sheriff brought me the body for identification. The dead man was still wearing most of his clothes; unfortunately, he wasn't still wearing his head. That would make it difficult, if not impossible, to identify him positively. "We need to find the head," I told the deputy. It might well be at the bottom of the Emory far from where the fisherman had found the body, but there was also a chance somebody had found the skull lying on the riverbank somewhere and maybe even picked it up.

The body arrived on a Wednesday. On Thursday the *Roane County News*, the area's weekly paper, ran a front-page story on the discovery of the body and the importance of the missing skull. The article asked anyone who had seen or acquired a skull to bring it to the sheriff's department. Over the next few days two skulls arrived, which deputies duly delivered to me.

The first one, which came in on Friday, was dry and dusty, clearly not from our recent, ripe floater. Two things about this skull intrigued me, though: the ethnicity, and the huge hole knocked in the base of the cranium. Our floater was Caucasoid, but this skull looked Japanese or Chinese, which made it an unusual find in East Tennessee. I asked the sheriff's office for the story behind it, and they told me the man who'd brought it in was a junkyard dealer. A few days before, he'd bought a junked car from a local landowner. Sitting inside a five-gallon paint bucket in the car's engine compartment was the skull.

The man who had sold the junker, it turns out, had served in the Pacific theater during World War II. While walking along a beach in Okinawa, he had stumbled across a crashed Japanese Zero; inside was the skull of the dead pilot, which our patriotic GI brought home as a war trophy. (In ensuing years I would encounter more World War II trophy skulls, almost all of them Japanese, almost none of them European in origin—an interesting commentary on our attitudes toward the dead from different cultures.) At some point between 1945 and 1973, the base of the Japanese pilot's skull—his foramen magnum—was knocked out so a light bulb could be inserted into his cranium: The dead warrior had been reduced to a mere Halloween decoration.

Skull number two was a Native American skull, also dry, dusty, and far older than our floater. The search for the missing skull would have to continue. Meanwhile, the unsolved mystery was starting to cause a stink—literally. Most cities have morgues where bodies can be kept in cold storage until they're identified and either claimed by relatives or buried by the local government. That's not the case in many small, rural towns—like Kingston, the seat of Roane County, where our floater had surfaced once enough decomposition gases had built up in his abdomen to make him buoyant. The deputy didn't want to take the smelly body back to Kingston with him, so I helpfully agreed to keep it at the university. The problem was, I didn't have a cooler, either. With the weekend now at hand, I wrapped the body in plastic, sealed it the best I could, and stashed it in the mop closet of a rest room near my office. I'm not sure how many people were in the building when the janitor came in to mop the halls that weekend, but I expect everyone who was—and probably a few passing motorists outside—heard him when he opened the stinky bundle in his closet and saw what was inside. On Monday morning he made it crystal clear—in language equally plain to a scientist or a sailor—that, department head or not, I was never, under any circumstances whatsoever, to store rotting bodies in his mop closet or anywhere else in his building. One single infraction, I gathered, might result in the subsequent discovery of my own headless body in very short order.

Ever quick to take a hint, I sought help from my boss, the college dean. I explained our little dilemma, which he comprehended with swiftness and equanimity. Opening a campus telephone directory, he thumbed through the listings for the College of Agriculture, made a brief call, and solved my problem: The ag school had several farms outside of town, and on one of those farms stood a vacant building, a sow barn, which was basically an open, three-sided shed. The farm's only neighbors were the prisoners in a county correctional facility, and they probably had better things to complain about than an occasional whiff of decomposition. It seemed like a good place to store bodies temporarily until we could clean and study the bones.

It worked fine for several years. Gradually, though, I started noticing something odd: Occasionally I would find a body in a slightly different position than the way I had left it a day or two before. I also noticed footprints and other signs of uninvited human visitors. Eventually we figured out what was happening. The convicts next door, working outdoors on the grounds of the penal farm, had discovered the sow barn's gruesome new residents and had taken to sightseeing. So far nothing had been removed, but I didn't want to take the chance of losing a crucial piece of forensic evidence—a skull containing a telltale bullet, for instance.

As I was mulling over the need for a new storage facility, along came Colonel Shy, telling me that merely warehousing bodies wasn't enough. I needed to do more than just remove the rotting flesh from bodies; I needed to *study* it, *observe* it, learn everything it could tell me about death and decomposition. I couldn't do that kind of research in a musty sow barn, especially not one located forty-five minutes away from my offices and labs. I needed someplace bigger, somewhere closer.

It was my sixth year as head of the anthropology department. By now our physical-anthropology faculty had expanded from one to three; our curriculum had grown from undergraduate courses to a full-fledged Ph.D. program; and we were beginning to attract some of the brightest and best graduate students in the nation. We had, in short, the resources to do something that had never been done before: to establish a research facility unlike any other in the world—a research facility that would systematically study human bodies by the dozens, ultimately by the hundreds; a laboratory where nature would be allowed to take its course with mortal flesh, under a variety of experimental conditions. At every step, scientists and graduate students would observe the processes, document variables such as temperature and humidity, and chart the timing of human decomposition. We would pick up where Sung Tz'u had left off seven centuries before.

The idea was simple; the implications—and the possible complications—were profound. By most cultural standards and values, such research could appear gruesome, disrespectful, even shocking. Yet the chancellor never questioned the wisdom of it; fortunately, he had watched and admired the growth of our program up until now, so he was unhesitating in his support. Once again it was a simple matter of a phone call.

Just across the Tennessee River from the main campus—barely a long punt from the football stadium, as the pigskin flies—was an acre or so of surplus land behind the UT Medical Center. For years the hospital's trash had been burned there, and it wasn't exactly prime real estate, but I'm not sure I'd have felt at home if it had been.

All my life I've scrimped and scratched and made do with very little. Growing up during the Great Depression, I saw how carefully my mother stretched the insurance money we got after my father's death. Excavating Indian graves on the plains of South Dakota, I fed crews of hungry college students on government-surplus peanut butter and berthed them on surplus Army cots. Moving into dilapidated quarters huddled beneath the football stadium—the windows looked out onto a maze of steel girders supporting the upper deck—I repainted peeling walls and refinished ancient dormitory desks and repaired hand-me-down filing cabinets. So when the chancellor offered me an acre of nearby land—even junky land—just five minutes from my office, I was grateful to get it: death's own acre, you could call it.

In the fall of 1980 my students and I set to work. We cleared trees and brush from the center of the site; we laid a gravel driveway so trucks could pull in with bodies and equipment; we ran a water line and electricity from the hospital. Working mostly by hand, we cleared and leveled a pad sixteen feet square beneath the shelter of the trees, then spread several inches of gravel. Once we had the sixteen-by-sixteen pad ready, I had a concrete truck come pour a load of concrete; together, the students and I smoothed its surface. Atop this slab we built a small frame building, simple and windowless, roofed with cheap asphalt shingles. The building would give us a place to store tools like shovels and rakes, instruments like scalpels and surgical scissors, and supplies like latex gloves and body bags. It ran the full width of the pad but extended only six feet deep. That left us a front porch, so to speak, measuring ten feet by sixteen feet. On it we could easily lay up to a dozen bodies for our decomposition studies.

The sow-barn visits by the convicts from the penal farm had shown me that security was important, so I decided we could afford, just barely, to fence our little square research area.

People who know about the Body Farm today seem to think it sprang into existence fully formed, but that's not the way it happened at all. It came from humble beginnings, and it progressed by small steps. The questions we hoped to answer were almost laughably elementary: At what point does the arm fall off? What causes that greasy black stain under decomposed bodies, and when? When do the teeth fall out of the skull? How long before a corpse becomes a skeleton? To find answers, we first had to find research subjects. We had the farm; now we needed the bodies. I sent letters to the medical examiners and funeral directors in Tennessee's ninety-five counties.

Finally, one Thursday evening in the middle of May, 1981, I drove a covered pickup truck to Burris Funeral Home in Crossville, Tennessee—an hour west of Knoxville, on the Cumberland Plateau—and picked up our first donated research subject. The corpse was a seventy-three-year-old white male who had suffered from chronic alcoholism, emphysema, and heart disease. We knew his identity—the body had been donated by his daughter—but for the sake of confidentiality, we assigned him a unique identifying number. In life he'd had a family and a name; in death he would be known simply as "1-81": the anthropology department's first donated body of 1981. (My forensic cases were identified by the same pair of numbers, but in reverse order: The first criminal case of that same year was case 81-1. The system wasn't fancy, but it worked.)

The following morning a handful of graduate students and I laid corpse 1-81 on the concrete pad we had poured a few months before. Someone took pictures. To protect 1-81 from rodents and other predators small enough to squeeze through the fence, we covered the body with a wooden framework screened with wire mesh. One by one we filed out of the chain-link enclosure. I closed the gate and snapped a padlock onto the latch. A fly brushed past my ear. The Anthropology Research Facility was embarking on its first research project. Death's acre was open for business. The Body Farm was born.

CHAPTER 8

A Bug for Research

O N A WARM, sunny day in 1981, as corpse 1-81 lay decomposing at my newly commissioned Anthropology Research Facility, almost visible across the river from the University of Tennessee's anthropology department, Bill Rodriguez and I stepped out from beneath Neyland Stadium. In Bill's hand was a glass vial containing five flies, and on the back of each fly was a dot of orange paint, bright as the jersey of a UT lineman.

Standing on the steps in the sunshine, Bill unscrewed the top of the vial. Within seconds all five flies were gone. We looked at each other and grinned. "Let me know what happens next," I said.

As it turned out, Bill was about to embark on a study that would help spur a revolution in forensic science, becoming one of the most heavily cited anthropology papers of all time. I didn't know that at the time, though.

At the time, all I knew was that there was a lot yet to learn about bodies and bugs.

I HAD MOVED to Knoxville ten years earlier, in 1971. I'd spent the sixties teaching in Kansas and excavating Indian graves in South Dakota; between the ancient Indian skeletons and the recent murder victims brought to me by local deputies and KBI agents, I'd seen somewhere in the neighborhood of five thousand bodies prior to coming to Tennessee. By then I figured I'd seen just about everything. I figured wrong.

During my first year in Knoxville, local and state police officers brought me around a dozen bodies to examine, and in at least half of those cases I found myself face-to-face with something I knew very little about: maggots.

Maggots are the small, wormlike larvae that hatch from the eggs laid on a body by flies—usually, but not always, the iridescent green insects called blowflies. When maggots first hatch, they're smaller than grains of rice; by the time they mature, they're roughly as long and fat as pieces of macaroni. They get that big by feasting on decaying flesh. In Tennessee they do, anyhow; in Kansas, not so much.

Kansas has a pretty dry climate, so bodies often mummify—dry out and shrivel up—before the maggots get to them. In Tennessee, on the other hand, there's twice as much rainfall and plenty of humidity between rainstorms; in summer you can almost steam broccoli just by setting it outdoors. All that moisture, plus all the shade from the Tennessee woods—there's not much prairie east of the Mississippi—tends to keep a corpse's flesh soft and easy for maggots to chew. It didn't take me long in Tennessee to learn to open body bags outdoors and on the ground, lest the morgue be overrun by maggots and flies.

I've had a strange, symbiotic relationship with flies ever since I was a small child. Shortly after my father's death, my mother and I moved in with her parents. We lived on a farm, and where there are farm animals, there are flies. My mother, who hated flies, made me a business proposition: for every ten dead flies I brought her, she'd pay me a bounty of one cent.

With an incentive like that, I became a six-year-old fly-killing machine. When Grandpa came in from milking the cows, I noticed, flies would flock to any drops of milk that sloshed out of his pail. *Swat*— seven with one blow! Before long I learned to cajole my grandmother for cups of milk, so I didn't have to wait for milking time or Grandpa's spills. The fly carcasses piled up, and so did my pennies.

Ever since, though—and as a scientist, I'm embarrassed to admit this—I have despised flies. I hate rattlesnakes more, but rattlesnakes are a lot less common, a lot more shy, and a lot easier to kill. As I'd learned in South Dakota, all it takes to decapitate a prairie rattler is a steady hand and a razor-sharp shovel. Flies, though, are relentless and almost infinite in number. Lay a fresh, bloody body out of the ground on a summer day, and within minutes the air will be thick with swarming blowflies. Swing a shovel like a giant flyswatter and you can probably knock down a few on the wing, but in the time it would take to do that, dozens of reinforcements will arrive.

Yet, watching the flies swarm, I knew there must be something that they and other insects could teach us. There had to be some way they could deepen our understanding of death—particularly the postmortem interval, or time since death.

I certainly wasn't the first scientist to notice how swiftly flies can sniff out the odor of death, how unerringly they are drawn to the scent of blood. Way back in A.D. 1247, the Chinese investigator Sung Tz'u recounted a murder case in his pioneering forensic handbook *The Washing Away of Wrongs*:

> There was an inquest on the body of a man killed by the roadside. . . . The inquest official familiarized himself with the victim's neighborhood. He thereupon sent a number of men separately to go and make proclamations. The nearest neighbors were to bring all their sickles, handing them in for examination.

If anyone concealed a sickle, they would be considered the murderer and would be thoroughly investigated. In a short time, seventy or eighty sickles were brought in. The inquest official had them laid on the ground. At the time the weather was hot. The flies flew about and gathered on one sickle. The inquest official pointed to the sickle and said, "Whose is this?" One man abruptly acknowledged it. . . . Then he was interrogated but still would not confess. The inquest official indicated the sickle and had the man look at it himself. "The sickles of the others in the crowd had no flies. Now, you have killed a man. There are traces of blood on the sickle, so the flies gather. How can this be concealed?" The bystanders were speechless, sighing with admiration. The murderer knocked his head on the ground and confessed.*

Six centuries later, in the 1890s, a New York entomologist named Murray G. Motter examined 150 bodies that were exhumed when a cemetery was relocated. Motter noticed that the bodies had fed and housed numerous species of insects, at various developmental stages (larvae, pupae, adults); ultimately some of the insects became entombed in the very corpses that had nourished them—an irony that probably went unnoticed and unappreciated by the bugs themselves.

Motter published his insect inventory in the *Journal of the New York Entomological Society* under the remarkably thorough title "A Contribution to the Study of the Fauna of the Grave: A Study of One Hundred and Fifty Disinterments, With Some Additional Experimental Observations." The study did not inspire other entomologists to follow in Motter's macabre footsteps, at least not with human subjects. However, sixty years later another entomologist—this one in Knoxville, by curious coincidence—made a detailed study of insect activity in dog carcasses. The Knoxville entomologist's name was H. B. Reed; the question that

* From *The Washing Away of Wrongs: Forensic Medicine in Thirteenth-Century China*, by Sung Tz'u, translated by Brian McKnight; © University of Michigan Center for Chinese Studies, 1981. Quoted by kind permission.

interested him was not forensic but ecological: How does a corpse alter the environment in the small ecosystem where it decays? To find out, over a one-year period Reed set out forty-five carcasses of dogs that had been euthanized by the local pound. He set out one every two weeks during hot weather; during cooler spells he lengthened that interval.

Reed made several intriguing observations. Not surprisingly, he found that the total number of insects on, in, and around the carcasses was greatest during the summer; however, several individual species experienced their population peaks during cooler weather. The woods were buggier, he noted, but decomposition proceeded faster in open areas—possibly because of higher temperatures, he theorized. Perhaps most important, Reed meticulously documented all the species of insects, both adults and larvae, associated with the dog carcasses.

In the 1960s a South Carolina entomologist named Jerry Payne made a similar study using baby pig carcasses. Payne's major contribution was his careful record of insect succession: that is, he noted who showed up, and when, to march in the insect parade.

Meanwhile, during my summers in South Dakota in the sixties, I had noticed an interesting phenomenon in the Arikara Indian remains I was excavating. Some of the graves contained numerous pupal casings—the hard, hollow shells in which maggots encase themselves for their metamorphosis into adult flies; other graves, though, contained few casings or none at all. Eventually the light dawned: During winter, flies are grounded by the cold; in fact, anytime the temperature drops below 50 degrees Fahrenheit, flies stop flying. The Arikara graves that contained no pupal casings held people who died and were buried during cooler seasons of the year. It fascinated me at the time to realize that we could figure out, two hundred years after it happened, what season of the year an Arikara warrior had fallen in battle. By the time I established the Body Farm, I knew that if I could get a graduate student interested in studying insect activity in corpses, we'd probably figure out ways to deduce a lot more than just the particular season in which a person had died.

Bill Rodriguez was the ideal graduate student for the task—partly because he was willing to take it on, and partly because he had a broader background in field research than most graduate students.

Bill had an undergraduate degree in anthropology, with a minor in zoology. He'd entered anthropology intending to study primates, and in fact he actually went to Africa as part of a team working to restore laboratory-raised chimpanzees to the wild. But he'd also taken my osteology course and had done quite well in it, so one day, when I needed someone to go with me on a forensic case, I went looking for an assistant, and Bill was the first qualified helper I found. He was washing grimy windows in one of our classrooms; because we were housed beneath the stadium's concrete stands, a lot of dust and dirt swirled onto and into our quarters. Bill had a teaching assistantship, which sounded pretty highbrow, but the "assistantship" part included some lowbrow chores like washing windows.

"I need somebody to go out on a case with me," I said. "Why don't you finish that later?" Bill was only too happy to oblige.

It was a cold, snowy day. The body, which had been discovered by a road crew picking up trash alongside a country road, was partially covered with mud. The skull lay ten feet or so away from the rest of the body; all the remains were largely skeletonized.

I asked Bill—as I always asked my students—to tell me what he made of the scene. He correctly identified the skull as a white male's; he also quickly determined that the man had been shot in the head. Then he pointed out what looked to be additional perimortem trauma to the skull and commented on the shallow burial.

His last two observations were logical but wrong. The marks he interpreted as trauma inflicted around the time of death were actually postmortem: they were tooth marks left by rodents (rats, probably) that had dragged off the skull and gnawed off bits of flesh. What appeared to be a shallow grave was actually an illusion: the body lay in a shallow creek bed that was dry when we were there; during rainy spells, though, muddy water had gradually deposited a thin layer of silt around and on top of the body.

The skull bore a couple of other interesting clues as well. The location of the gunshot entry wound—just behind the right ear, with a fracture pattern suggesting that the barrel of the gun had been pressed against the skull—marked this as an execution-style killing. One of the zygomatic bones, or cheekbones, was deformed in a way I'd seen several times before. It had been broken, probably in a bar fight—and probably by a pool cue, judging by what I'd learned about several prior victims who had almost the exact same pattern of trauma and healing. His teeth had several unfilled cavities and lots of chewing-tobacco stains, so clearly he was not exactly from the upper crust.

As we excavated, we noticed numerous pupal casings in and around the remains; that told me that—like those Arikara Indians whose graves had first gotten me thinking about insects—he'd been killed during warm weather. The vines and roots growing under parts of the body tended to confirm that as well.

The police never managed to solve that particular murder, but the case did have one happy ending: it got Bill Rodriguez hooked on forensics. Primatology lost a promising young scientist that cold, snowy day. Not long after that, Bill helped clear the ground, level the gravel pad, and pour the concrete for the new Anthropology Research Facility. A few months later he helped me set out our first research subject, corpse 1-81. By then Bill had settled on his thesis topic. H. B. Reed had chronicled insect activity in dog carcasses. Bill would do the same thing with human corpses, beginning with 1-81.

THE INSECT STUDY was not a pleasant project. Besides 1-81, we'd brought over a decomposing body from the sow barn; in addition, over the next few months, we acquired another couple of bodies.

Bill put the bodies up on wire racks, so he could observe and gather insects from beneath them. Then he parked himself on a stool for hours every day and watched what happened.

What he saw first, with each of his four experimental subjects, was a profusion of blowflies. The warm-weather bodies, like 1-81, began attracting blowflies by the hundreds within a matter of minutes. Blood triggered a feeding frenzy like nothing he could have imagined: Sitting just a foot or two away from a bloody body, Bill would soon find even himself overrun with flies, seeking any moist bodily fluids to feed on, any dark, damp orifices (including Bill's nostrils) to lay their eggs in. He quickly learned to wrap netting around his head to keep the flies out of his eyes, nose, mouth, and ears.

On a warm day it took only a matter of hours for the nose, mouth, and eyes to be filled with grainy, yellowish-white masses of fly eggs. One female blowfly can lay hundreds of eggs at a time, and there were literally thousands of pregnant females swarming around each body after its arrival. In the heat of May and June—the months when 1-81 and 2-81 were placed in the research enclosure—those clumps of eggs hatched into thousands of maggots in as little as four to six hours.

But the flies weren't the only bugs to flock to a fresh body. Yellowjackets and wasps showed up within the first minutes to hours too. Some of them fed on the body itself, Bill noticed; others snagged flies on the wing, carried them off, and decapitated them with one swift bite of their jaws. Still others feasted on the masses of fly eggs or the tender young maggots hatching in the body's openings.

As the maggot population exploded, Bill noticed carrion beetles arriving to feed not just on the carrion but on the maggots as well. Like a wasp beheading a fly, a beetle would clamp its powerful jaws on a wriggling victim and cut it cleanly in two. Bill described some of these life-and-death struggles for me in epic terms; I don't know that I've ever seen a student so thoroughly immersed in a research project. "This is the food chain in action," he told me excitedly one day. "This isn't just some happenstance occurrence; it's an orderly sequence, it's something we can interpret and use forensically."

Bill's research was a breath of fresh air for the field of anthropology, but not for his home life. After a day parked on his stool, surrounded by bodies and buzzing insects—many of which would light on him after feeding on corpses, and some of which would even lay eggs on him—

he'd go home with the reek of decomp on his clothes, his skin, his hair. After the first day or two Bill's wife, Karleen, issued strict orders: He was to strip in the garage, put his clothes straight in the washer, and jump into the shower immediately. Then, and only then, was he permitted to approach her.

Early in the study—just a matter of days into it—Bill and I were speculating about how far away the flies could smell the bodies, and whether the same flies were coming back day after day to feed on them. That's when we got the idea of marking the flies with orange paint and trying to track them.

Using the net with which he gathered specimens every day, Bill caught five blowflies buzzing around corpse 1-81. He brought them back to my office in the department and painted the thorax of each one with UT orange so they'd be easy to spot in a swarm. When we took the marked flies outside and released them, they took off, seemingly at random. The next day at the Body Farm, though, Bill netted three of the five marked flies.

ON FEBRUARY 11, 1982, nine months after the study began, Bill presented his results at the annual meeting of the American Academy of Forensic Sciences in Orlando, Florida. The room, a large banquet room in a big Hyatt Hotel, was fairly crowded as Bill got up. Within minutes, though, as he began projecting the 35mm slides he'd taken at frequent intervals during the study, people began to get up and leave the room. Were Bill's slides—the first images we'd shown of human bodies decomposing at the research facility—too disturbing for even seasoned forensic scientists to stomach?

Another few minutes passed, and the people who had left the room began returning—accompanied by throngs of others, summoned from other presentations scheduled simultaneously with Bill's. "You've *got* to come see this" was the message that spread like wildfire through the Hyatt's meeting rooms that day.

Bill went on to publish his results in the *Journal of Forensic Sciences* that fall, and that article, "Insect Activity and Its Relationship to Decay Rates of Human Cadavers in East Tennessee," became one of the most cited, most reprinted articles in the journal's history. In fact, in a 1998 brochure highlighting the fiftieth anniversary of the American Academy of Forensic Sciences, Bill's talk got mentioned as one of the organization's high points—"the first of the 'bug' papers," the brochure called it.

As one of the rising young stars of forensic anthropology, Bill took some interesting jobs after graduate school, including positions with a forensic consulting laboratory in Louisiana and with the medical examiner in Syracuse, New York. His most unusual job, though, is his current one: He's the staff forensic anthropologist for the Armed Forces medical examiner, whose office is responsible for identifying and, when needed, autopsying the bodies of military personnel, diplomats, spies, space shuttle astronauts, and anyone else sent by the federal government—or nearby state and local governments—for examination.

IN APRIL OF 1986, while he was still working for the Louisiana forensic lab, Bill was asked by police in Falls Church, Virginia, to examine evidence gathered from a death scene a year and a half earlier.

In August of 1984, Lisa Rinker, an eighteen-year-old girl, had left her house one Sunday night around 10:30, telling her mother she was going for a walk around the block. She never returned, and the next morning her mother contacted the police to report her missing. Police, family, and friends began searching the town and surrounding areas, but they found nothing.

The following Saturday night, one of Lisa's friends—the best friend of Lisa's boyfriend, in fact—brought Lisa's father a pair of familiar-looking pink flip-flops, which he said he'd spotted at an intersection outside town as he was putting up Missing posters. Her sister, Nancy, confirmed that the flip-flops were Lisa's.

Mr. Rinker rounded up a group of relatives and friends, and the next day they headed out to search the woods near the intersection. The searchers set about their work with a sense of grim foreboding, for the smell of death was in the air, and strong. About sixty yards from the highway guardrail they found Lisa's body lying in dense underbrush. She was wearing dark blue corduroy jeans with white trim on the pockets—the same pants she'd been wearing the night she disappeared—and a shredded tube top. Her torso was covered with maggots; her face had been eaten away, and so had her internal organs. The skin on her hands and feet had begun to slip, or slough off. Her feet were bare; however, despite the rough terrain and dense underbrush, the soles showed no traces of bruises or scratches. The lack of trauma to the soles, together with a difference in skin color around the toes and arches, suggested that she'd been wearing something on her feet at the time of her death, and possibly for some time afterward as well.

Two days later the local medical examiner performed an autopsy. Because of the advanced state of decomposition and partial skeletonization, he was unable to tell what had killed her. He listed Lisa's cause of death as undetermined, and her grieving parents buried her.

But the police investigators weren't ready to put the case to rest. Lisa had quarreled bitterly with her boyfriend, Bernie Woody, on the night she disappeared. According to police, Lisa had been cheating on him—with her own sister's husband, Dale Robinson—and witnesses had told investigators the boy had threatened her. A car owned by his friend Danny Heath, the fellow who spotted Lisa's sandals beside the highway, was reported parked beside the road that night near the spot where Lisa's body was eventually found.

The detectives leaned hard on Lisa's boyfriend and his pal Danny. During a polygraph test, a police statement said, Danny appeared to be lying when he was asked questions about Lisa's death. With no cause of death and nothing but circumstantial evidence to suggest that Lisa might have been murdered, though, the district attorney decided not to file criminal charges against either Bernie Woody or Danny Heath.

Meanwhile, a new investigator, Rick Daniele, had become fascinated with the case. Daniele sent photos of the body to Dr. Louise Robbins, a forensic anthropologist in North Carolina, along with the flip-flops found beside the highway. Dr. Robbins, an expert in footprint and shoe print analysis, told Daniele that the discoloration patterns in the forefoot and arch areas indicated that the flip-flops had remained on her feet for several days after she was killed. Dr. Robbins also noticed a piece of sloughed-off skin stuck to one of the flip-flops—further proof that the body was partially decomposed when the sandal was removed.

That's when Detective Daniele contacted Bill Rodriguez and asked him to analyze the evidence. Besides the photos, he sent Bill soil samples that had been gathered at the death scene, along with preserved maggots that had been collected from Lisa's body. It was obvious that the investigators had done a thorough job of gathering evidence; it was less obvious, but just as significant, that entomology had become a respected forensic tool, thanks in great measure to Bill's insect study at the Body Farm five years before.

As Bill leafed through the photos of Lisa's body, he was struck immediately by the advanced state of decomposition, particularly in the chest region and the hands. Lisa's face was completely gone, but that wasn't too surprising: with its moist openings, the face is a blowfly's preferred place to feed and lay eggs—usually, that is. But not when there's blood somewhere else on the body.

Any forensic anthropologist who's seen a victim who's been stabbed to death or whose throat has been cut knows how dramatically the presence of blood at the sites of those wounds attracts flies and promotes maggot growth. Within days, if the weather is warm—as it was in August of 1984, when Lisa Rinker died—the masses of maggots hatching in the bloody wounds consume the surrounding tissues far faster than they otherwise would. It's a phenomenon we call "differential decomposition," which raises a red flag instantly in the mind of any trained forensic scientist.

From the extent of differential decomposition in her chest and abdomen, Bill was virtually sure Lisa had been stabbed there; the damage to the soft tissue of her hands suggested that she'd been cut there, too, probably trying to defend herself. He called Detective Daniele to tell him so.

Armed with Bill's reading of the photos, Daniele retrieved Lisa's clothing from the evidence file and sent it to the Virginia crime lab. The crime lab's analysis bore out Bill's hunch: tests of eight stained areas of Lisa's pants indicated the presence of blood—lots of blood, enough to have soaked the fabric. Daniele pleaded with the family and the district attorney to allow Lisa's body to be exhumed so Bill could examine it for signs of skeletal trauma.

Three months later, on a cold, snowy day in January, Bill arrived at the cemetery where Lisa had been buried. Breaking through the frozen ground, cemetery workers unearthed her coffin and hoisted it out of the ground; then they put it into a hearse for the trip to the Fairfax County morgue. There, Bill removed the chest, abdomen, and both hands, placed them in a large kettle of water, and boiled them for an hour to remove the flesh. Then he removed the bones from the kettle and gently brushed them clean.

Lisa Rinker had indeed been stabbed. Bill found a total of seven knife marks—several in different parts of the chest cavity (the ribs and sternum), plus defensive wounds on both hands. The knife marks were made by a thin-bladed knife, Bill's examination found. According to police, Danny Heath often wore a large pocketknife in a sheath on his belt, but they say after Lisa's murder, he stopped wearing it. The cause of death on Lisa's death certificate was changed: *Undetermined* was struck through, and *Homicide* put in its place.

Sadly, Lisa's killer remains at large. Despite the skeletal proof Bill found showing that Lisa had been murdered, and despite the lingering questions surrounding Bernie Woody and Danny Heath, the Fairfax County Commonwealth Attorney remains unwilling to prosecute the case.

Anthropologists and insects can reveal the truth about a crime, but they can't force the wheels of bureaucracy to turn, and they can't guarantee that justice will be done. All they can do is serve as a voice for victims, and hope that voice is heard.

CHAPTER 9

Progress and Protest

O N MAY 15, 1981, when we laid my first research subject at the Body Farm, corpse 1-81, in the sixteen-foot-square chain-link enclosure that was the Anthropology Research Facility, the daytime high was just 58 degrees. Over the next few days, though, temperatures shot up into the eighties. A couple of months earlier and we might as well have put him in a meat locker, but once the hot weather hit, the changes were swift and dramatic. Within days the flesh of the face was nearly gone, consumed by the maggots hatching in the mouth, nose, eyes, and ears. Bill Rodriguez was carefully charting the insect activity, but the changes in the body itself, and their timing, were fascinating—and gruesome.

There are four broad stages in a body's decomposition: the fresh stage, the bloated stage, the decay stage, and the dry stage. Some scientists tend to split these into finer gradations, but I try not to get bogged down by definitions. (There are two kinds of observers in science: splitters and lumpers. I've never been much of a splitter; in my heart of hearts, I'm a lumper.)

In 1-81's fresh stage, the body's toothless upper jaw and yellow-toothed mandible stretched what was left of a face into a kind of grin. As the insects multiplied and fed, they soon left gaping eye sockets to stare at us blindly. The hair and skin retained their hold on the skull, but within days their grip was clearly beginning to slip.

By the end of the first week, the corpse began to bloat. As bacteria began to consume the stomach and intestines, the abdomen started inflating almost like a balloon from the waste gases of the microbes. Meanwhile, the skin began turning a rich, reddish brown. Fatty tissues began to break down beneath the skin, giving the corpse a glossy shine, almost as if it had been basted with a glaze and baked in an oven.

As the flesh turned the color of caramel, a network of purplish-crimson lines began to show through it, like a satellite map of a continent's rivers. We were seeing the circulatory system, its veins and arteries highlighted as the blood within them began to putrefy, making them larger and darker, almost as if they'd been outlined on the body with a felt-tip marker.

The graduate students and I watched in utter fascination. As far as I knew, no scientist had ever done this before: deliberately set out a human body to decompose, then simply sat back and watched, taking systematic note of what happened and when. Many scientists—and even the artist Michelangelo—had studied bodies, but their focus was human anatomy; by dissecting the dead, they hoped to learn more about the flesh and bone of the living. My interest was death itself.

Two weeks into 1-81's journey from fresh corpse to bare skeleton, his skull was bare bone. The hair had slid off in a mat, still held tenuously together by tangles and a bit of tissue. The hair mat lay in a dark, greasy pool of goo that surrounded the head. His bloated abdomen had collapsed, leaving his belly shrunken and clinging to the jutting rib cage, marking his transition from the bloated stage to the decay stage. Within another week, the ribs themselves—along with the vertebrae of the spine—were exposed. So were the bones of his pelvis, as the result of a vigorous insect assault on his genitals and the surrounding area.

His limbs had decomposed more slowly. Lacking the moist, dark openings of the face and pelvis, the arms and legs were less desirable territory to the insects colonizing the body. One dramatic and fascinating change had taken place at the hands and feet, though: About seven days into the process, the skin began to soften and slough off in large sheets, almost as if 1-81 had suffered a particularly severe sunburn and his skin were peeling. At first the sloughed-off skin was pale and pliable; amazingly, the ridges and whorls of the finger- and toe prints were still clearly visible, a fact I relayed to one of my friends at the Knoxville Police Department, Arthur Bohanan, who was KPD's top fingerprint expert. Within a few days, though, the skin had dried and shriveled, almost as dead leaves do. But when Art took one of these shriveled husks back to his lab, he managed to moisten and uncurl it, coaxing 1-81's identity once again from something an untrained investigator might well have discarded as leaf litter.

One month after his arrival, 1-81 was little more than a skeleton. Some leathery skin remained on his rib cage and skull, where the sun had dried or mummified it to the texture of leather; beneath it, though, all his soft tissue had been consumed by bacterial action and insects. I left his bones to bleach for four or five months, then gathered them up and brought them into the hospital morgue for "processing"—cleaning off the last vestiges of dried skin and cartilage. Then I measured the bones, recording the key dimensions: femoral length; femoral head diameter; cranial length, breadth, and height; the distance between the eye orbits; and a host of other data that would preserve the measure of the man.

The skeletal measurements were part of a larger plan that had been taking shape in my mind over the preceding months and years: to build the largest collection of skeletons—*modern* skeletons—in the United

States. Several immense skeletal collections already existed. The Terry Collection, originally housed in Saint Louis at Washington University but later sent to the Smithsonian Institution, included more than 1,700 individual skeletons; the Smithsonian's other collections, as I knew from personal experience, possessed far more, including thousands I had sent there during my summers excavating in South Dakota. But those bones were old, and for forensic purposes that made them obsolete.

In many ways we humans have taken ourselves out of the evolutionary loop. Take me, for example. I'm terribly nearsighted; my vision is about 20/200. If I'd lived ten thousand years ago, I wouldn't have survived long enough to reproduce and pass along my myopia; squinting hard, I might have glimpsed the saber-toothed tiger about the time he opened up his jaws to chomp down on my neck. Today, fit or unfit for the rigors of "Nature, red in tooth and claw," we survive and we breed. (Of my three sons, two—Jim and Charlie—inherited my nearsightedness; my middle son, Billy, somehow ended up with eyes sharp enough to qualify him as an Army helicopter pilot.)

But appearances notwithstanding, we continue to evolve, including our skeletons. A century ago the average white American male stood five feet seven inches tall; today he stands five feet nine inches. The average Arikara Indian female measured five feet three inches back in 1806, when Lewis and Clark glimpsed her standing on the bank of the Missouri River; today she's two to three inches taller.

When an unknown crime victim is found—especially if police find only a few of the long bones—the only way to estimate stature accurately is to compare those long bones to the average dimensions of corresponding bones from individuals of known stature. And if the numbers being used for comparison are out of date, the estimation could be off by several inches. As a result, instead of searching for a missing male six feet tall, the police might mistakenly search for a missing male five feet nine. Data from 1-81 could help prevent such mistakes.

One other way in which 1-81 would continue to help us for years to come was as a teaching tool. Learning the size, shape, and feel of every bone in the human body is an enormous challenge for anthropology students. The only way to do it is to study actual bones—real ones, not plastic or plaster casts of them—for countless hours. In my osteology class every semester, the students used to dread the "black box" test: I'd put several bones inside a black box that had circular openings cut into the sides; to pass the test, a student would have to reach in and tell me, just by feeling them, what bones (or, if I was feeling merciless that day, what bone fragments) were in the black box. Even something as subtle as weight and texture can be crucially important. The skulls of blacks, for instance, are denser, heavier, and smoother than the skulls of whites; that's one key reason there have been so few outstanding black Olympic swimmers: they have to work harder just to stay afloat. In a forensic case, if only part of a skull is found, knowing the difference in density and heft could help tell police whether the victim was white or black.

Our donated body, 1-81, had died of disease, but my plan was to build a skeletal collection that would include victims of trauma too. That way, when I lectured about antemortem and perimortem fractures, students could see how bones broken before death had healed, while bones fractured at the time of death didn't. When I described gunshot entry wounds and exit wounds, students could see and feel how the entry fracture tends to bevel, or widen at an angle, as the bullet penetrates the skull; what lead spatter looks like on the inside of the cranium; how much larger an exit wound is and how it, too, bevels and widens in the direction of the bullet's travel.

Much of our early research focused simply on observing and recording the basic progression and timing of decomposition. As Colonel Shy had made painfully clear, our understanding of postmortem processes was quite limited. The questions these studies hoped to answer were simple, but the answers would take years to piece together. Every variable made a difference: Was the body in sunlight or shade? Clothed or nude? Outdoors, or in a building—or a car? The passenger compartment or the trunk? On land or in water? One early experiment posed a deceptively simple question: How far away can the smell of death be detected by the human nose?

As usual, it was a real-world case that set me to thinking about that question. This one happened right in my own backyard—or almost. The backyard where it happened was only a few miles north of the anthropology department's offices and labs, off a busy thoroughfare named Broadway. Technically it wasn't a backyard but a vacant lot between a house and Broadway, covered with weeds, brush, trash, and piles of dirt. In the summer of 1976, the owner of one of the adjoining houses finally got tired of looking at the mess, so he called the property's owner to complain. The owner obligingly hired a cleanup crew, which brought over a tractor equipped with a front-end loader to scoop up the trash and brush.

Several hours and truckloads of debris later, as they got close to the center of the lot, one of the workers spotted what looked like a human skull lying in the weeds. He called his buddies over to confer, and they agreed with his skeletal analysis. Needless to say, that was the end of the cleanup work for the day. The workers called the police, and the police called me.

I headed out Broadway, accompanied by Pat Willey, the graduate student who ran the osteology laboratory—my bone lab. Pat and I did a little digging and found a few more bones, but not many. Most of them, we soon realized, had probably already been scooped up and hauled off to the landfill.

From the condition of the bones—they were completely dry and sun-bleached—it was immediately apparent that they'd been lying in the lot for quite some time, possibly several years. It didn't take long to make a positive identification, either: The top plate of the dentures was prominently labeled *Orval King*, the name of a local man last seen some two years before. A seventy-four-year-old who'd spent some time in the regional psychiatric hospital, he had either fallen or lain down in the vacant lot between a house and a busy street and quietly died.

In this case, the tantalizing puzzle to be solved was not who he was, or how long he'd been dead, or even why he died. This time what baffled me was why he hadn't been found shortly after he died. More precisely, why hadn't he been *smelled* shortly after he died? When an adult human male decomposes, the smell can be overpowering, as you can well imagine if you've ever driven slowly past a dead dog with your car windows open on a warm summer day.

We knew that the house adjoining the vacant lot was occupied at the time the man died; we also knew that the sidewalk across the front of the lot carried a lot of neighborhood pedestrians, and Broadway was one of Knoxville's busiest streets. Yet no one had smelled anything, or at least nothing bad enough to prompt suspicion, investigation, or complaints to the city.

So if the stench of death didn't carry as far as the houses or the sidewalk, how far *did* it carry? Or, put another way, if the human nose couldn't detect a body at that distance, at what distance *could* it detect a decomposing corpse? The answer would be useful not just to me, I figured, but to police, firefighters, and search-and-rescue workers almost anywhere.

Orval King had raised an intriguing research question. Now, at our new two-acre research facility, I had the perfect place to determine the answer scientifically, experimentally. All I needed was a dead body and some live guinea pigs.

The body arrived soon enough: an unclaimed corpse from a nearby medical examiner. The guinea pigs? A cinch. Undergraduates will do *anything* for extra credit. To recruit volunteers for this experiment, I announced in my fall Anthropology 101 class one Thursday that anyone who wanted to earn ten extra points should meet me at the research facility on Saturday morning. The turnout was incredible. Nearly a hundred students crawled out of bed early on a weekend—all of them motivated, I'm sure, by selfless scientific fervor.

The experiment was simplicity itself: I had positioned a body, which was extremely bloated and quite smelly, a ways up the gravel road leading into the facility. The body was hidden from view by trees and bushes.

The day before, I'd put markers at ten-yard intervals from the body—that is, a marker at ten yards, twenty yards, thirty, forty, and fifty. Then I led my student guinea pigs up the primrose path one by one. "Tell me when you smell something" was the only instruction I gave; then, on a clipboard I carried, I made a slash mark in the columns corresponding to the distance each student indicated. As I led them toward the corpse they would begin inhaling sharply and concentrating intensely. Most of them wouldn't say anything until we were twenty or even ten yards from the body, then they would wrinkle up their noses and say, "Phew, something really stinks here."

The research was quick and dirty, as we say in academia. It wasn't the sort of thing I'd ever write up and publish in the *Journal of Forensic Sciences*, but it was good enough to show me that yes, you *can* die and decompose in a vacant lot between a house and Broadway and never be smelled by thousands of people passing by just fifty feet away.

OUR FIRST SEVERAL YEARS of research were a time of exciting progress. Bodies had begun arriving almost weekly from medical examiners and donors. In fact, not only was the concrete pad inside our chain-link enclosure filled to capacity, but we'd added three additional shelves—bunk beds for the dead—up the sides of the fence.*

I surveyed our expanding research program with eagerness and pride. It's true, what they say: Pride goeth before a fall. One spring day in 1985, I arrived to find half of my two-acre research fiefdom flagged off with surveyor's stakes. To one side a bulldozer idled ominously. I buttonholed one of the surveyors and asked what was going on. The hospital parking lot was being expanded, he told me. As it turned out, the ag school had given me more land than it actually owned; instead of a two-acre former dump, I actually possessed a one-acre former

* For more information about UT's forensic anthropology program—including body donation—see the program's website, www.utk.edu/~anthrop/FACcenter.html.

dump, and no appeals on my part could stop the bulldozers and graders and pavers.

But losing half my land turned out to be the least of my worries. A few days later I got called out of a class—a drastic measure, practically unprecedented—by Annette, the departmental secretary. Did I know about the protest out at the Body Farm? I did not. Annette and I jumped into a car and drove over to the hospital parking lot, where we parked in a distant, inconspicuous corner.

A local health-care advocacy group called Solutions to Issues of Concern to Knoxvillians—S.I.C.K. for short—was picketing my research facility. Draped across one side of the fence was a giant banner that proclaimed, "This makes us S.I.C.K.!" Even though my facility was the target of the protest, I couldn't help laughing when I saw the sign. It was smart, it was funny, and it got them great news coverage.

But what had brought the wrath of S.I.C.K. down upon me? It seems that one of the survey crew laying out the parking-lot expansion had taken his sack lunch into the shade one day and suddenly found himself staring at the bodies decomposing inside our little chain-link enclosure. He went home and complained to his mother, who just happened to be one of the leaders of S.I.C.K. As any concerned mother would, she quickly organized a protest.

When I explained the purpose of the facility—researching decomposition to help police solve murders—the group acknowledged that yes, such work had scientific merit, but why did it have to be located here, practically under the public's nose? Couldn't we move it, say, twenty miles west, onto the vast, wooded, and heavily guarded government reservation at Oak Ridge?

Well, hell, I'd just moved the damned thing from twenty miles away barely a year before; one of the keys to establishing our research program had been finding a location close to the anthropology department. I phoned the university chancellor, Jack Reese, and explained the dilemma. I didn't want to cause any trouble for UT, but I sure would hate to lose or relocate my research facility. Jack was as wise as Solomon and as generous as Carnegie. He offered to pay, out of his own budget, for the installation of a chain-link fence around our remaining acre of woods, to keep people from wandering close to the bodies.

A few weeks later the fences were up and the crisis was over. Robert Frost was right: Good fences *do* make good neighbors. But it wouldn't be our last crisis—and it wouldn't be our worst.

CHAPTER 10

Fat Sam and Cadillac Joe

I GOT A CALL one Thursday in May that made me close my office door. That was a rare thing. I kept my door open pretty much all the time—partly because I liked to see what was going on in the department; partly so students and faculty would feel free to talk to me about any little problem they were having (before it became a big problem); partly so nobody would wonder or worry or gossip about what was going on behind Dr. Bass's closed door. So when they heard my phone ring and my door close, everyone in Anthropology figured something sensitive was going on. It was.

The call came from Arzo Carson, the director of the Tennessee Bureau of Investigation. The TBI and the FBI were working together, he said, on a case that had begun as a kidnapping but had apparently escalated into a murder. Carson didn't need to tell me that, with the FBI looking over his shoulder, the stakes and the pressure were sky-high for the TBI.

As curious grad students tiptoed past my door, straining to catch snatches of the conversation, Director Carson briefed me on the case. The circumstances—hell, even just the criminals' names—were the most bizarre I had ever encountered in a forensic case: Fat Sam. Cadillac Joe. Funky Don.

After I hung up, I opened the door and called in two of my best forensic response team regulars, Pat Willey and Steve Symes. Without giving them any details, I asked if they wanted to help me with some fieldwork the next week. Steve and Pat both agreed instantly, clearly eager to pierce the veil of mystery. Five days after the TBI director's call, the three of us piled into my station wagon and headed west on I-40 toward Nashville. As we drove I filled them in on the case.

Fourteen months before, a couple named Monty and Liz Hudson were kidnapped in broad daylight from the parking lot of a Nashville hotel. The hotel, a Holiday Inn, was in a fairly safe part of town, adjacent to the campus of Vanderbilt University. In plain sight of several witnesses—including one with a camera, who took photographs—the Hudsons were abducted at gunpoint by three men. Two of the kidnappers forced Monty Hudson into his own Cadillac, the third shoved Liz into another car, and the two cars left the Holiday Inn together.

A couple of days later Liz Hudson was released in downtown Nashville. By then the kidnapping had been reported, and agents from the TBI and FBI were crawling all over the parking lot and the Holiday Inn for clues. That's when the case started to get really strange.

Liz refused to cooperate with the FBI. She told them that the kidnapping had been a simple misunderstanding and that Monty had since left town on a business trip. She didn't know where he'd gone or when he'd be back, but she assured them that Monty was fine and nothing was amiss. Liz was six months pregnant at the time of the kidnapping. Three months later she gave birth to Monty's baby, but Monty still wasn't back from that business trip.

A couple more months went by. Then the investigators got a tip about Monty's whereabouts: According to an informant, Monty's business trip had ended in a shallow grave some seventy-five miles south of Nashville, on a farm near the Alabama border.

WEST TENNESSEE is cotton territory. Nashville is music territory. Lawrence County, in 1980, was "Fat Sam" Passarrella's territory. Think of mobsters and you'll probably picture wise guys from Jersey or Chicago or Vegas. The town of Lawrenceburg, Tennessee, most likely doesn't spring to mind in connection with organized crime, but it should. Well, maybe not *organized* crime, actually; more like *disorganized* crime.

Fat Sam hadn't always been called that. His mama had christened him Sam John, but that had been many years and about four hundred pounds ago. Sam grew up in New York, but apparently he fell in with a bad crowd up there, so his family sent him down south to straighten him out. His aunt Louise owned the local telephone company in Lawrenceburg and was a pillar of the community; under her positive influence, the family hoped Sam would embark on a business career of his own.

He did. By 1980, Sam's numerous business ventures included counterfeiting, money laundering, marijuana farming, drug distribution, and trafficking in stolen property. This smorgasbord of illegal enterprise had caught the eye of a joint FBI–TBI–Secret Service task force on organized crime, and the task force was amassing a fat file on Fat Sam and his cohorts: "Funky Don" Parsons, Howard "Big Daddy" Turner, Elvin "Bank Robber" (sometimes shortened to "B.R.") Haddock, and Earl (no nickname) Carroll.

In the months following Monty Hudson's disappearance, the task force began tightening its net around Fat Sam's gang. When Sam was indicted for counterfeiting, the others could see the handwriting on their indictments as well. One of them, Earl Carroll—perhaps figuring the first to squeal would get the best deal—contacted a Nashville FBI agent, Richard Knudsen, and offered to spill the beans on Fat Sam's crimes, including, he claimed, the kidnapping and murder of Monty Hudson.

Carroll spun a wild tale. Monty Hudson was a con man, he said, nicknamed "Cadillac Joe" because of his fondness for stealing that particular brand of car. But cars weren't the only hot properties Monty had acquired. According to Carroll, Monty contacted Fat Sam and offered to sell him a batch of pure silver bars, more than thirty in all, each measuring a good foot and a half long by about six inches wide and four inches high. Tipping the scales at nearly one hundred pounds apiece, each was stamped with a mint mark and serial number attesting to its authenticity. At the time, silver was selling for up to $50 an ounce—about ten times what it is today. At those prices just one of Monty's silver bars could be worth up to $80,000. But because he needed to sell them quickly and discreetly, no questions asked, he was willing to make Sam a hell of a deal on them: $20,000 cash would buy the whole lot of them.

Fat Sam was interested, but he wasn't so gullible as to take Monty's story on faith. One of his cronies, Funky Don, had some experience with precious metals, and Fat Sam asked Funky Don to run a test, or assay, on one of the bars. He did, and it assayed as pure silver. Sam forked over the twenty grand, and Monty handed over the silver. But as Fat Sam discovered when he had it retested, it wasn't silver after all; it was actually zinc, another soft, heavy, silvery metal but worth only a few cents per ounce. In other words, for his $20,000, Fat Sam had bought a batch of metal bricks worth less than a hundred bucks. Sam was furious, Carroll told the FBI: furious at Funky Don—maybe he'd botched the assay, or maybe he was in on the scam—and even more furious at Monty.

That's when he swooped down on Monty and Liz in the parking lot, just as they were about to skip town. At some point after the kidnapping, Liz was being held elsewhere while Fat Sam and Big Daddy Turner, who was actually a small man, took Monty for another ride in his Cadillac. Sitting in the backseat, Monty said something smart-alecky. It was the last thing he ever said: one of the two men up front—it's not clear which one—turned around and shot him.

Now there was the problem of Liz, Monty's wife: She hadn't seen the murder, but she could certainly link the men to the kidnapping. Fat Sam didn't have the heart to kill her, so he called in a hard case, an out-of-towner from across the border in Alabama. Apparently the hired killer took one look at Liz—a beautiful woman, by all accounts, and obviously pregnant to boot—and announced, "No matter what kind of a sorry son of a bitch I am, I can't kill no pregnant woman." At that point, said Carroll, Fat Sam released Liz and ordered his cronies to dig two graves in remote areas outside Lawrenceburg: one for Monty, and one for . . . his Cadillac!

I've heard some pretty outlandish stories over the years, but Earl Carroll's took the cake. Apparently the FBI and TBI believed it, though, because it wasn't too long after he told it that I found myself headed toward Nashville in search of Monty Hudson. Along with me were Steve, Pat, and an assortment of shovels, trowels, wire-mesh screens, and evidence bags.

We met FBI Agent Knudsen, several TBI agents, and a state prosecutor for breakfast at a Shoney's on the south side of Nashville, then piled into their cars for the trip to Fat Sam's territory. The agents were visibly nervous, so the idea of including a professor's station wagon in the convoy seemed like a risk to them. We headed south on I-65 for an hour or so, then got off at the exit for Pulaski, another small town near the Alabama border. There, in a Wal-Mart parking lot, we picked up yet another TBI agent: Bill Coleman, based in Lawrenceburg, who was the TBI's point man, or "case agent," investigating Fat Sam's activities.

After picking up Coleman in Pulaski (the birthplace of the Ku Klux Klan, by the way), we headed out into the country. In the course of about ten miles, we went from a four-lane U.S. highway to a two-lane blacktop to gravel to dirt. The dirt track, an old logging road, ended in a clearing that was being swiftly reclaimed by honeysuckle vines, blackberry bushes, and tree seedlings.

The instant the cars jounced to a stop, the FBI and TBI agents jumped out, guns drawn in case we were ambushed by Fat Sam and his gang. For once I wished I'd taken TBI Director Carson up on his offer to issue me a TBI firearm when he presented me with my TBI consultant's badge. I actually went out to the firing range and shot well enough to qualify once—at night, to boot—but then I decided that it was silly for me to carry a gun. For one thing, by the time I get called to a crime scene, I'm a lot more likely to be confronting dead victims than live criminals. For another, I'm generally in no position to defend myself anyway, crawling around with my nose to the ground and my rear end in the air.

In this case, my rear guard looked pretty competent: half a dozen or so armed state and federal agents, swiftly fanning out around the clearing to establish a secure perimeter. The absence of sheriff's deputies at a rural scene like this was unusual; the organized-crime task force suspected that some of the local lawmen were not to be trusted, I later learned from Bill Coleman. The TBI and FBI wanted us to arrive unannounced and, if possible, undetected. Me, I was just hoping we'd be able to depart unharmed.

The FBI agent, Knudsen, had been here once before, led by Earl Carroll. According to Knudsen, Carroll walked to a spot about fifty feet to the left of the logging road, looked down, and started cussing. "Well, this is where he *was*," he'd told Knudsen, pointing at a shallow trench in the ground where he said that he and another of Fat Sam's cronies had buried the body.

Knudsen led me to the spot in question. It was choked with weeds, briers, bushes, and poison-oak vines, but even so, I could tell at a glance that the ground had been disturbed fairly recently. Atop the area of disturbed earth, a log and several tree branches had been laid side by side. Mixed with the reddish-brown clay was a white, powdery material, which Carroll had told Knudsen was lime, dumped over Monty Hudson's body in a misguided effort to speed its decomposition. (That seems to be a common misconception among murderers. Lime does reduce the odor of decomposition, but it also reduces the *rate* of decomposition. As a result, a lime-covered body may be less likely to get sniffed out, but it's more likely to linger.)

As a TBI agent videotaped the proceedings, we set to work. First, Steve Symes photographed the scene from several viewpoints, starting from beside the cars, then gradually working his way closer. Then Pat Willey and I began clearing away the brush, vines, and grass. Even before we started to dig, we made a key find. Lying in a tangle of weeds and leaves and small rocks was a human ulna from a right forearm.

Whoever had moved the body—Fat Sam or his henchmen—had done a pretty sloppy job of it, but that wasn't surprising. Put yourself in the shoes of the body-mover and you'll see why: You go out to dig up a body and hide it someplace else. This body, mind you, has been decomposing in a shallow grave for months now, so it's going to be really smelly and mighty rotten. You hold your breath, grab an arm, and give a pull . . . and the arm comes off in your hands. At this point, unless you're exceptionally conscientious and have an iron stomach, what you're going to do is scoop up whatever big pieces you can grab between breaths of fresh air—a head, a torso, a couple legs, most of the arms—and then hightail it out of there as fast as you can. Fortunately for me, most bad guys sent to move a rotten body don't know or don't care that teeth can fall out after a few weeks, hands can drop off or get gnawed off, bullets can work free and get left behind.

Since the grave appeared shallow, we excavated with trowels rather than shovels. After a couple of hours of careful digging, we had dug down to the undisturbed layer of earth. By then we'd found a jumble of other things besides the ulna: two thoracic (upper back) vertebrae. Fifteen teeth. Four fragments of an occipital, the shattered base of a skull. Five finger and toe bones. A fragment of a long bone, possibly from a tibia (shin). Human hair. Empty pupal casings left behind by maggots metamorphosing into adult flies. Tatters of cloth. A bullet.

We bagged the teeth and bones to take back to the anthropology department for a thorough examination, and we gave the cloth and the bullet to the TBI for analysis. Clambering back into the government cars, we headed back to Nashville, then went our separate ways, safe and sound.

Back in Knoxville, we began sifting through the material we had in order to determine the Big Four: sex, age, race, and stature. Unfortunately, we didn't have a whole hell of a lot to go on. Determining the sex was complicated by the lack of a pubic bone, hipbone, or face. However, the ulna was massive, and that strongly suggested that the sex was male. So did the fragments from the occipital: the external occipital protuberance—the bump at the base of the skull—was quite pronounced and bore heavy muscle markings, characteristic of a man's neck muscles.

The age was more difficult to pin down, since all we had to judge by was the presence of osteoarthritic lipping. The ulna showed some early ("first stage") lipping at the elbow joint; so did the finger and toe bones and the thoracic vertebrae. That meant he was probably somewhere between thirty and fifty years old—so maybe somewhere around forty—but it was impossible to be any more precise than that.

Without a face or cranial vault, determining the victim's race was tough too. The hair was dark in color and badly matted; from a simple visual examination, we couldn't determine the victim's race. We set aside a sample for more detailed study later.

We were in better shape to determine his stature. We had one long bone, an ulna, and from its length we could extrapolate to estimate the victim's height. There was one complication: the distal (lower) end of the ulna had been chewed off by a carnivore of some sort, so we first had to figure out how long the bone had originally been before it got gnawed down to about 29.5 centimeters. By comparing it with several complete ulnae, we determined that less than 5 percent of the bone was missing; that meant the complete bone would have measured about 31 centimeters. Plugging that number into a formula developed by anthropologist Mildred Trotter back in the 1950s gave us an estimated stature of around six feet one to six feet two.

Our studies of decomposition and time since death were just getting under way at the Body Farm in 1981, so we had little research data to compare with what we observed in the remains we'd recovered from the field. Bits of dry tissue remained on some of the bones; the odor of decay was pronounced but not overpowering; and numerous empty pupal cases were interspersed with the bones. On the basis of my observations of other decayed bodies over the prior twenty-five years, I put the time since death at somewhere between one and three years.

The teeth, I hoped, might be the key to telling us whether or not this was part of Monty Hudson's body we'd found. Of the fifteen teeth we'd found, seven—nearly half—had fillings, some of them fairly large and distinctive. If we could lay our hands on Monty's dental X rays—assuming they existed—we should be able to tell pretty quickly whether Earl Carroll's story was true.

By this time the FBI had told Liz Hudson that Monty was probably dead, and she agreed to help in any way she could. Her earlier silence had been motivated by the best of intentions: She didn't know Monty was already dead by the time she was released in Nashville, and she desperately hoped that by keeping quiet, she was keeping him alive. A little naïve, maybe, but also deeply loyal and very brave. Now Liz told Agent Knudsen everything she remembered about the kidnapping and began suggesting where to ask for dental records.

Monty had lived for quite a while in Tulsa, she said, so Knudsen began contacting dentists there. He struck pay dirt pretty soon: Dr. R. Jack Wadlin confirmed that Monty Hudson had been a patient of his, and he agreed to send dental charts and four bite-wing X rays of Monty's teeth. The fillings and the pulp cavities, the internal structures, shown in the X rays from Dr. Wadlin matched the fillings and the X rays we took of the teeth we'd recovered from the shallow grave in rural Tennessee. It was indeed Monty Hudson—a little bit of him, anyway—that we'd excavated.

In the months following our field trip to Fat Sam's territory, he and his two partners stood trial for the kidnapping of Monty and Liz Hudson. Big Daddy Turner was also charged with Monty's murder. All three men were convicted on both counts of kidnapping. By then, Passarrella was already facing a stiff sentence for counterfeiting; the kidnapping sentence tacked on another twenty years. I hear Fat Sam's gotten religion while serving his time, as well as become an accomplished gardener or amateur botanist. I also hear his nickname still fits pretty well.

Big Daddy Turner ended up taking the worst fall. Offered a sentence of just two years if he pled guilty to lesser offenses, he turned it down, choosing to take his chances with a jury trial. The gamble cost him dearly: he was sentenced to forty years for the kidnappings—twice as much as Fat Sam—plus life in prison for felony murder. After a series of appeals, eventually he pled guilty to two counts of aggravated kidnapping and to "accessory before the fact" to second-degree murder, but he still drew concurrent forty-five-year sentences for the three crimes. Turner had chosen what was behind Door No. 2, you might say, and what was behind it turned out to be a set of steel bars, a whole lot of years, and Turner himself. Meanwhile, just as he'd hoped, Earl Carroll—the stool pigeon—got off the lightest. I read in the newspaper he got a sentence of just two to ten years. My friends in law enforcement tell me he's been back in at least once since, but is currently on the straight-and-narrow road, literally, as a truck driver.

Monty's Cadillac, it later emerged, was buried several miles away, in a field where Fat Sam subsequently planted a large marijuana crop. The TBI had raided the field and destroyed the plants; by amazing coincidence, TBI agent Bill Coleman had sat on a mound of earth as he watched the crop's destruction—the very mound of earth bulldozed atop the Cadillac. After it was unearthed, the car was hauled to the TBI crime lab outside Nashville. Fat Sam needn't have gone to the trouble of burying it: the lab technicians found no bloodstains or other incriminating evidence anywhere inside.

Where the rest of Monty's body ended up, I've never heard. The story goes that after Earl and B.R. had buried Monty in the shallow grave, Fat Sam went out to inspect their handiwork and found it wanting; apparently the body was almost entirely exposed. As the old saying goes, if you want something done right, do it yourself. Fat Sam wasn't as thorough a grave-robber as he might have been, but he was certainly better at holding his tongue than Earl Carroll was.

Thirty-one of the "silver" bars that set the killing in motion were eventually dredged up from a creek bed in rural Giles County, not many miles from the site of Monty's initial grave. They were right where Earl Carroll said they'd been dumped. TBI agent Bill Coleman, now retired, has hung on to one of them as a souvenir. Liz Hudson, Monty's beautiful widow, settled down in Nashville, went to work for one of the city's many music-related companies, and settled down with a country-music songwriter. Somehow that seems fitting. Any day now I expect to turn on the radio and hear a woeful ballad about Fat Sam and Cadillac Joe. If it ever plays out that way, Monty might finally make that fortune after all—not quite the way he intended, but maybe on a far bigger scale: Through the alchemy of country music, those zinc bars of his could someday be turned into gold or even platinum. I suspect he'd like that.

CHAPTER 11

Grounded in Science

I NEVER CEASE to be amazed by the whys and the ways human beings commit murders—and by the new techniques forensic scientists develop to solve these crimes. Some of those techniques, I'm proud to say, are being devised by people I've trained.

On September 20, 1991, I got a call from Jim Moore, a TBI agent based in Crossville, a small city about sixty miles west of Knoxville. Some bones, possibly human, had been found in the crawl space beneath a house outside Crossville. Agent Moore wondered if I could come over the next day with a forensic response team to excavate the bones and determine whether they were indeed human.

Unfortunately, I couldn't go, I told him: I was leaving early in the morning for Washington, D.C., to teach a forensic anthropology class at the Smithsonian Institution for medical examiners from around the country and for agents from the Smithsonian's next-door neighbor, the FBI. I could, however, send over an experienced forensic response team.

By this time the forensic response teams worked like a well-oiled machine, even without me. I rounded up the graduate students who were on call—Bill Grant, Samantha Hurst, and Bruce Wayne—and relayed Agent Moore's instructions: They were to meet him at his office in the Cumberland County courthouse in Crossville at 12:30 the next day, then follow him out to the scene. As they left my office I gave them one final reminder: *"Don't forget Arpad's soil samples!"* A revolutionary new technique for determining time since death was about to get its first test in a murder case.

During the decade since we'd begun studying human decomposition at our research facility, we'd done dozens of studies and experiments, most of them involving the many variables that affect the rate of decomposition. We'd seen bodies hold together throughout winter and much of spring, and we'd watched them skeletonize in as little as two weeks during the muggy heat of summer. We'd compared bodies tucked in the shade with bodies baked in the sun and found that the bodies in sun tended to mummify, their skin becoming tough as leather, impervious to maggots. We'd compared bodies on land to bodies submerged in water; the floaters lasted twice as long. We'd compared bodies on the surface to bodies buried in graves, ranging from shallow to deep; the deeply buried bodies took eight times as long to decompose as the exposed bodies. We'd compared fat bodies to skinny ones; the fat ones skeletonized far faster, because their flesh could feed vast armies of maggots; in fact, one recent follow-up study, measuring daily weight loss in cadavers, recorded an astonishing forty-pound weight loss by an obese body in just twenty-four hours—a record I'm sure no fad diet will ever come close to.

All of these studies shed important light on the events and timing of human decay, but they all relied on an observer's interpretation of visible, gross changes. (By *gross*, I mean the changes were large-scale.) So although we'd made every effort to detail and differentiate those changes as thoroughly as possible, there remained room for subjective interpretation and, therefore, an element of imprecision. Determining time since death was still a frustrating, inexact science.

Then, a few years after we'd begun our research, a young scientist approached me with the audacious and ambitious proposal to *make* it an exact science. His name was Arpad Vass, and he worked in a commercial laboratory that analyzed forensic specimens for law enforcement agencies. Arpad proposed entering our Ph.D. program and developing a quantitative, scientific technique that would rely on biochemical data to determine time since death. In effect, he proposed inventing a forensic clock that could be run backward, starting the moment a body was found. When it stopped—when it unwound all the way back to zero, basically—it would tell the time of a murder victim's death.

Arpad had a bachelor's degree in biology, with a minor in chemistry, and a master's degree in forensic science—great credentials for a criminalist. But Arpad wanted to do more than work in a crime lab: he wanted to advance the frontiers of forensic technology. The idea was fascinating. If it worked, it would offer a revolutionary new way—a quantitative, objective way—to answer one of the first and most crucial questions every homicide detective asks: How long has this person been dead?

I had two nagging concerns about Arpad's proposal. First, how in the world could we define a chemistry project as anthropology research? Second, and far more crucial, could he make the technique work?

I've always been a big believer in the cross-fertilization of ideas. Every forensic investigation is a team effort, and the more experience—the more *kinds* of experience—the better, in my opinion. Not all of my colleagues in the field share that view; while I've improvised down in the bowels of a football stadium, some anthropologists have dwelled high in the proverbial ivory tower, looking down their noses at our unorthodox methods in Tennessee. But over the years I've noticed that my knowledge as an anthropologist has been greatly enriched by things I've learned from people who came to this field by an unconventional route.

Take Emily Craig, for example. Unlike our typical graduate student, she hadn't come to us waving a freshly inked B.S. in anthropology; in fact, she was in her forties by the time she applied to our Ph.D. program. Emily had a master's degree in medical illustration, and she'd worked for years in a Georgia orthopedic clinic, illustrating scientific articles and guides to surgical procedures. In the course of that career she'd spent a lot of time around doctors and seen a lot of bones, so I figured she might bring an interesting perspective to her anthropology studies. As it turned out, I was wrong—by way of underestimation, that is.

Her first semester, Emily took my human-identification class, in which students learned how to look at skeletal remains and determine the Big Four: sex, age, race, and stature. I brought in one skeleton every other week—a known skeleton, often one from a forensic case the police had brought to me.

About six weeks into the course—about the time the students were starting to get cocky—I'd always throw them a curve. An elderly black man had wandered away from a nursing home in Winchester, Tennessee, years before; when a skeleton was eventually found, the authorities asked me to determine if this was the missing man. I didn't think so, I told them initially: The skull wasn't Negroid; the teeth and jaws didn't jut forward the way a black man's would. Pat Willey, the graduate student who ran my bone lab at the time, agreed with me. Then, a week later, we received X rays of the missing black man—which matched the skeleton we'd confidently pronounced as white.

Every year in Human Identification, I led my students down the same primrose path I'd taken with that skeleton, and invariably the students—noting the absence of prognathism in the mouth structure—would write *Caucasoid* on their test paper just as confidently as I had pronounced it years before.

When I got to Emily's paper I was shocked: *Negroid*, she'd written—

the only one in the class to get it right; the only one *ever* to get it right. I called her into my office and confronted her. "Tell me who told you that was a Negroid skeleton," I demanded. For years I'd been fooling students with this trick question, then sworn the class to secrecy afterward, so the next year's students would likewise learn not to jump to conclusions so quickly. Now, it seemed, somebody had broken the code of silence.

"No one told me," she said. She sounded surprised and indignant.

I persisted: "Then how did you know? *Everybody* gets that wrong. They take one look at that skull and they're sure it's Caucasoid."

"I didn't look at the skull," she answered. "I looked at the knee."

I stared at her, utterly baffled. "What on earth are you talking about?"

My student then proceeded to explain to her professor—a diplomate of the American Board of Forensic Anthropologists—that the knees of blacks have more space between the condyles—the broad, curved ends of bone that form the knee's hinge—than the knees of whites. "That's why surgeons would much rather operate on the knees of black athletes than white athletes. There's a lot more room to work in. *Everybody* in sports medicine knows this."

At this point I was more than three decades into my career, yet this was a complete revelation. "*Nobody* in anthropology knows this," I told her. After I swallowed a few mouthfuls of humble pie and collected my professorial wits, I added, "This would make a great dissertation topic."

Emily took my advice. Not only did she research, confirm, and publish what she'd already noticed in the knees of living athletes, she went a step further: Another subtle difference in the knees of blacks, she found, could be used to estimate race in unidentified bodies. The angle of an interior seam in the femur just above the knee—called Blumensaat's line, in honor of the German physician who first noticed it in lateral (side-view) X rays—differed in whites and blacks. After taking hundreds of X rays and measurements of femurs, Emily devised a formula that could distinguish, with up to 90 percent accuracy, a Negroid femur from a Caucasoid femur. In a field that had previously depended solely on the skull to determine race, this was a remarkable advance.

If Emily hadn't come to anthropology by way of medical illustration, we might never have learned about this, and we'd have missed out on a technique that has proven crucial in identifying several unknown murder victims.

It was that same sort of scientific cross-pollination that Arpad Vass was proposing in his plan to use biochemical data to pinpoint time since death. In his case, though, it wasn't bone structure he was talking about, but bacteria.

As Arpad talked about turning bacteria into a forensic stopwatch, I tried to think of some other department where his research might fit better than in anthropology. I knew it was too applied and too forensic-oriented to gain approval in the biology or chemistry departments. I also figured it would be a stretch to admit him into the anthropology program. But I couldn't stop thinking about how the field might benefit from such a revolutionary technique. "Tell you what," I finally said. "I'll fight to get you in, if you'll definitely tie it to human decomposition—and if you're sure you can make it work." He assured me that he would and he could.

It didn't take long to show me that he was serious about the first requirement. Within a matter of days Arpad was out at the research facility, taking samples of decaying flesh, maggot soup, and greasy soil. He'd gather a batch of samples, disappear into a chemistry lab for days, then reemerge to gather more goo.

It was that second part of our deal—making it work—that would be the hard part. Arpad had theorized that as a body decayed, a succession of different bacteria would feed on the decaying tissues, in the same way that a consistent succession of insects did. "Pigs is pigs," an old saying goes; Arpad's hope was that bugs is bugs, whether those bugs are macroscopic or microscopic.

In theory, his idea was simple. In reality, though, it was overwhelming. Looking at the samples under a microscope was like looking at an aerial photograph of Saint Peter's Square during the Pope's annual Easter sermon: the field of view was packed with individuals of seemingly endless variety.

He didn't tell me at the time, but Arpad spent months at the microscope, staring and despairing. It would take an immense laboratory, with a staff of perhaps fifty, to identify and track the legions of microbes converging on his research subjects, digesting their tissues, and leaving behind greasy puddles of waste. Then it hit him: the microbes themselves might be too difficult to analyze, but the grease slicks they left behind—the by-products and waste products created by the digestion of soft tissues—might contain some useful evidence.

Arpad took another look at his samples—not at the bugs themselves this time, but at the smelly soup in which they swam. Chemically the liquid around and beneath decaying bodies proved to be a mixture of various compounds, mainly volatile (lightweight, easily evaporated) fatty acids created by the breakdown of fat and DNA. As he studied the samples he'd collected over the weeks and months, Arpad realized that the ratio of compounds continuously evolved as the bodies decomposed further and further. In other words, a sample taken from beneath body A five days after death would differ markedly from a sample taken fifty days after death. Arpad really began to get excited when he noticed that the same patterns or ratios—the same evolving chemical profile—that held true for body A also held true for body B, body C, and so on.

By then Arpad knew he was on the trail of a consistent scientific phenomenon he could measure and harness. All he had to do now was track the ratios over time, then develop a procedure for taking a sample from a crime scene, determining the ratio of volatile fatty acids in that sample, adjusting for the average daily temperatures, then comparing that ratio with the ratios he'd observed at known postmortem periods. Oh, and develop a formula or equation that could easily calculate time since death, by matching his crime scene ratios with the ratios he'd carefully measured during two years of research at the Body Farm.

It's a difficult concept to explain—heck, it's a difficult concept for me to understand, not being a chemist—but an simple analogy might make it a little easier. Suppose you know that Joe Blow eats a scrambled egg every morning for breakfast; sometimes he also chops up a boiled egg in a can of tuna for lunch; and if he's feeling really ambitious, he might whip up a batch of chocolate-chip cookies using another two eggs. Now, if for some reason you happen to rummage around in Joe's garbage can, you should be able to tell, from the ratio of eggshells to tuna cans and chocolate-chip bags, how many days' worth of Joe's garbage you took out of that can.

What, you may be wondering, does all this have to do with some bones—possibly human bones—buried under a house in Crossville, Tennessee? Quite a bit, I hoped, which is why I wanted to be sure the forensic response team remembered to bring back soil samples.

The house belonged to a man named Terry Ramsburg. But Terry Ramsburg wasn't around; in fact, nobody had seen hide nor hair of him in more than two years, including his wife, Lillie Mae.

Actually, by now Lillie Mae was his ex-wife. She had reported Terry missing on January 16, 1989. He'd left the house for work at his auto body shop one day, she said, and didn't come home that night. When he still hadn't come back a week or so later, she finally called the police.

Not too long after she reported him missing, Lillie Mae filed for divorce, on the grounds that Terry had deserted her. In due course the divorce came through, and Lillie Mae subsequently remarried. She stayed in the house, just in case Terry should happen to resurface, and her new husband moved in with her and her two daughters.

Terry's father, Robert Ramsburg, didn't quite believe Lillie Mae's story. He knew things had been stormy at home—Terry expected Lillie's teenaged daughters to help out at the body shop, and they didn't like that—but he didn't believe Terry would simply leave town without a word. And when Lillie Mae got married again, Robert got more suspicious. His mind kept coming back to that house, and eventually he decided to snoop around a bit. One September day when nobody was home, Robert opened the wooden door leading to the crawl space. Holding a flashlight in one hand, he scuttled around beneath the floor joists, looking for something—anything—that might tell him about his son's disappearance.

In the far corner of the crawl space, he found it: a bit of red cloth protruding slightly from the soil. It was in a patch of dirt that seemed disturbed, fluffier than the hard-packed clay beneath most of the house. He tugged gently, exposing more fabric; then, using his bare hands, he began to scrape away the dirt. Gradually the red fabric assumed the familiar outline of a pair of long johns, and then he saw, jutting from the waistband, something that looked like bone. At once he stopped digging, went inside, and called the sheriff's office. A few phone calls and hours later, my graduate students were on their way.

For years our forensic response teams had carried essentially the same set of tools: shovels, trowels, rakes, paper evidence bags, plastic body bags, wire-mesh screens, cameras. Now, they took a small but significant addition: a pair of Ziploc plastic bags in which to collect soil samples—one sample from beneath the remains, another from an uncontaminated region ten feet away.

Agent Moore was waiting at the courthouse. So was Lillie Mae, who had consented to the search. They drove the mile and a half to the house caravan-style, with the white UT truck trailing the TBI sedan and Lillie Mae's car. With his characteristic thoroughness, Bill Grant jotted down her license tag: RNW-016. Several other cars were already parked at the house. Some had delivered a handful of city police and sheriff's deputies, but sitting quietly in one car was a pair of civilian onlookers: Terry Ramsburg's father and mother. Lillie Mae kept her distance.

Bill, Samantha, and Bruce quickly gathered their tools and crawled beneath the house. Agent Moore had already set up a work light in the crawl space, so the area was well lit. It took only a glance for Bill to confirm that the exposed bone was an innominate—a hipbone—and that it was human. Bill crawled to the doorway, extricated himself, and walked over to the small knot of officers. Robert Ramsburg got out of his car and joined the group; Lillie Mae edged over too.

"It's definitely human," Bill said. Terry's father hung his head. Lillie Mae spun on her heel and strode away.

"This is bullshit," she snarled. "This is fucking bullshit." She got into her car, slammed the door, and cranked the starter.

Bill looked at Jim Moore and asked, as tactfully as he could manage, "Are you sure you want to let her leave?"

Moore looked unruffled. "She's not going anywhere," he said, with the assurance of a lawman who knew how to assess someone's risk of flight.

Bill crawled back under the house, and the forensic team got back to work. As the most seasoned member, Bill was in charge. He put Samantha to work excavating the legs and Bruce exposing the left side while he moved up to the spot where he expected to find the skull.

In just a few minutes of troweling Bill found the back of the skull, indicating that the body was lying facedown. Toward the right side of it was a small, neat hole, its edges beveled so that it was slightly larger on the inside than on the outside. A fracture ran from the top of the hole all the way across the skull toward the left side. "Looks like we've got a gunshot entry wound," he told Samm and Bruce.

Bill gently troweled the earth away to reveal the skull without moving it, an excavation technique called "pedestaling." As he exposed the left side of the head, he saw more fractures near the forehead—a web of bone fragments angling outward—but no hole. "Hey, guys, the bullet might still be in the cranium," he said excitedly. A few minutes more and Bill had completely exposed the skull. The cartilage joining it to the cervical vertebrae had long since decomposed, so Bill reached down and lifted it. As he rotated it to look at the face, he heard a small clatter within the cranium: a .22-caliber bullet rattling around in the space created by the drying and shrinking of the brain.

The sobering reality of the situation hit them hard when they'd finished excavating the remains, gathering the soil samples, and boxing everything up for the trip back to Knoxville. They put the remains, the clothing, and the soil samples in a cardboard specimen box, measuring one foot square by three feet long. As Samantha emerged from the crawl space carrying the box, Robert Ramsburg started toward her. In a panic she turned to Bill Grant. "What do I do?" she whispered. "Is he going to want to see the remains?"

"This is evidence," said Bill. "He can't. Don't say anything; don't even look at him."

Eyes on the ground, Samm walked to the truck. From her downcast gaze and stricken face, Robert Ramsburg had to have a pretty good idea what the box contained.

He was right. It was his son.

It came as no surprise to anyone involved in the case that the anthropological examination indicated that the skeletal remains were those of a white male, age twenty-eight to thirty-four, measuring five feet five inches to five feet ten inches tall. It was also no surprise that dental X rays confirmed the victim to be Terry Ramsburg, a thirty-three-year-old white male who stood five feet six inches tall before a bullet laid him low.

I sent copies of our forensic report to the TBI, the Cumberland County sheriff, the Crossville Police Department, and the district attorney's office on October 9. That same day, Lillie Mae Ramsburg Davis was charged with first-degree murder and held without bond.

Her trial was set for July of 1992. For months she proclaimed her innocence. Then, a week before the trial was scheduled to start, Lillie Mae cut a deal, pleading guilty to second-degree murder. Investigators told me that she'd shot him on the sofa as he lay sleeping, then dragged him under the house and buried him. Shockingly, she continued to inhabit the house, along with her two daughters, directly over Terry Ramsburg's decaying body for another two and a half years; for part of that time her new husband lived atop the remains of his murdered predecessor.

Lillie Mae was sentenced to thirty years, but she became eligible for parole in just ten. At her parole hearing Robert Ramsburg, her former father-in-law, testified passionately against her release, and the parole board voted to keep her in prison.

Lillie Mae's guilty plea turned time since death into a moot issue, legally speaking, but scientifically it was still important. Terry Ramsburg's body had largely skeletonized under the house, except for a large quantity of adipocere beneath the chest and abdominal regions. (Adipocere—literally, "grave wax"—is a soapy, greasy substance that forms when fat decomposes in a damp environment.) The extent of skeletonization and adipocere formation told me that Terry Ramsburg had been in that crawl space a long, long time, probably since the day of his murder.

Could Arpad's soil analysis confirm or pinpoint the time since death with any greater precision? Well, as often happens with new scientific techniques, in this case we learned more about the technique itself than about the murder to which it was applied.

All the volatile fatty acids Arpad tested for were below detectable limits, and those limits were pretty doggone low: twenty-two parts per million. In plain language, the body had lain there so long that the flesh-eating bugs had long since moved on to greener pastures, and even their waste products had evaporated into thin air. Temperature measurements taken in the crawl space suggested that the body could have reached that point in about eleven months, whereas nearly three times that many months had actually elapsed since his disappearance. The technique, we now realized, was better suited to bodies that were still actively decomposing.

After the Ramsburg case, Arpad Vass continued to refine his soil-analysis technique for estimating time since death. He also developed other ways to harness cutting-edge chemistry to catch killers. Recently he devised a similar technique that analyzes tiny tissue samples from a murder victim's liver, kidneys, brain, or other organs. If the body is no more than a few weeks old, this tissue-biopsy technique can pinpoint time since death to within a matter of days or even hours. Now Arpad is working to isolate and identify the specific molecules that constitute the distinctive odor of death—the molecules that cadaver dogs respond to—as a step toward developing portable systems that police and human-rights investigators could use to locate clandestine graves.

And Arpad's original breakthrough—analyzing soil samples to determine time since death—has proved its accuracy and value in dozens of cases. The investigation into one of those cases would begin just three months after Lillie Mae confessed to shooting Terry Ramsburg and burying him under the house. Time since death—and Arpad—would play a prominent role in the "Zoo Man" murders.

CHAPTER 12

The Zoo Man Murders

EVERY OCTOBER, the East Tennessee hills get all gussied up for six dazzling weeks. The dogwoods turn crimson; the maples, a brilliant red-orange; the tulip poplars, bright yellow; the oaks, variations in red and brown.

Nine miles east of downtown Knoxville, not far beyond a bridge where Interstate 40 crosses the green waters of the Holston River, the fall colors put on a show in a thick stand of hardwoods paralleling the highway. The woods lie at the end of a short dead-end road, Cahaba Lane, which runs for a half-mile beside the eastbound lanes of the freeway. Facing the traffic are a handful of houses and trailers and a church perched high on a grassy slope, East Sunnyview Baptist Church. To the south, away from the interstate, a small wet-weather creek winds through the trees.

Cahaba Lane dead-ends at the foot of a towering billboard—*Comfort Inn, Free Breakfast, Guest Laundry*—supported by five rusting I-beams. Between two of the supports, a path leads up a gently sloping ridge that is dotted with empty beer cans, snack wrappers, egg cartons, shoes, and other household and automotive debris. The forest floor is also littered with acorns, which support a large population of squirrels.

On October 20, 1992, a hunter—aiming to do a little squirrel population control—wandered up the path into the woods. A ways up the trail he noticed a battered mattress and a rotting doghouse; stuffed into the doghouse was a department-store mannequin. Kicking aside some of the debris, he saw that the "mannequin" was actually a young woman: a chemically blond, partially nude, and very dead young woman, her hands bound with orange baling twine. The hunter raced to a phone and called the police. Within minutes the dead end began filling with vehicles from the Knox County Sheriff's Department and the Knoxville Police Department. One of the KPD officers who converged on Cahaba Lane recognized the victim as Patricia Anderson, a thirty-two-year-old white female he'd been trying to find since she disappeared nearly a week before.

Patty Anderson was no stranger to the police. A prostitute with a cocaine habit and a police record, she was good-looking and a flashy dresser. She was also in the early stages of pregnancy, a fact that few of her coworkers or clients knew. She'd told a bail bondsman that she was trying to scrape together enough money for an abortion; her quest for cash was probably what had brought her to this unlucky dead end.

The Knox County medical examiner quickly confirmed what officers at the scene had suspected from Anderson's battered face, bruised neck, bulging eyes, and livid face: After tying her up, someone had beaten and strangled her. Ironically, hundreds of people must have been passing by just a stone's throw away; if she called for help, her cries might have been drowned out by the constant roar of the traffic.

Anderson had last been seen on October 13; the next day her

boyfriend spotted his car—a Chevy Malibu, which she had taken—parked at a motel frequently worked by Knoxville prostitutes. But by then she had vanished. To officers familiar with the city's seamy underbelly, a murder suspect sprang immediately to mind when her battered body was found. He liked to rough up streetwalkers, and he'd done that at Cahaba Lane at least twice before. The hunt was on for Zoo Man.

Eight months before Patty Anderson was killed—back on February 27—a Knoxville prostitute had called the police to report that a "john" had hired her and driven her out to Cahaba Lane. Once there, she said, he took her into the woods and proceeded to rob, rape, and beat her. Then, in the middle of winter, he left her tied up in the woods, naked. She managed to free herself, get to a phone at a beauty shop nearby, and call the police.

A Knoxville Police Department investigator, Tom Pressley, drove the woman back to Cahaba Lane later that day so he could inspect the scene. An aging Buick LeSabre was parked at the end of the road. "That's it! That's his car!" the woman exclaimed.

Pressley parked and headed into the woods, accompanied by the woman. About a hundred yards up the path, the woman began to tremble. Clutching Pressley's arm, she pointed and whispered, "There he is now!" The scene was shocking: A man was standing in the woods with his pants down around his knees; in front of him, on her knees, was a sobbing woman. The officer drew his gun and approached, unnoticed.

Pressley ordered the man to lie facedown in the woods. Then he cuffed him, led him back to his squad car, and radioed for backup. One of the officers who responded drove the two women back to town; Pressley took the man in and booked him.

The man caught with his pants down was Thomas Dee Huskey, age thirty-two; he lived in a trailer with his parents in Pigeon Forge, a small town twenty-five miles east of Knoxville. Huskey was charged with rape and robbery. (A wallet belonging to the woman who had led Pressley out to Cahaba Lane was found in the floorboard of the LeSabre.) But a grand jury dismissed the first woman's complaints; the second woman left town and never showed up to testify against him. After several months in jail, Tom Huskey was released.

A couple of weeks after his release, Huskey was picked up again, this time for soliciting an undercover policewoman for sex. He was cited and fined, then released again. But he remained a prisoner of lust and rage, which he continued to direct at prostitutes. Among the streetwalkers, he soon acquired a bad reputation and a memorable nickname: "Zoo Man." He'd worked for two years at the Knoxville Zoo as an elephant handler, until he was fired in 1990 for abusing the animals. His job wasn't the only reason for the nickname, though: Both during and after the time he worked at the zoo, Huskey liked to take prostitutes to an empty livestock barn beside the zoo. There, rumor had it, he liked to tie women up and abuse them. By the summer of 1992 the word had spread among Knoxville's prostitutes: Stay away from Zoo Man.

Not everyone got the message, though. One Sunday afternoon in September, Huskey picked up another prostitute and took her out to Cahaba Lane, promising her $75—nearly twice her usual fee. But once they were in the woods, she later told police, Huskey tied her hands behind her back, then beat her and raped her. As he'd done to his victim in February, he left her tied up on the ground.

Just a few weeks later, on the night after Anderson's body was found, the police arrested Tom Huskey in Pigeon Forge, at the trailer he shared with his parents on Huskey Lane. Searching the trailer, they found a piece of orange baling twine in Huskey's bedroom—the same kind they'd found tied around Patty Anderson's wrists. They also found an earring, later identified as hers; snagged on the earring was a blond hair. Lacking a follicle, or root, the hair itself didn't contain enough DNA to compare with the victim's. However, a chemical analysis by the FBI crime lab showed that the hair found in Huskey's bedroom had been dyed with the same dye as Patty Anderson's hair.

The next step in the quest for evidence was to search the two places

Huskey was known to take women for sex: the barn by the Knoxville Zoo and the woods off Cahaba Lane. Six or eight local prostitutes had gone missing over the past few months, and if Huskey had killed one of them, as the evidence sure seemed to indicate, maybe he'd killed others, too.

Of course, just because a prostitute drops from sight, that doesn't mean she's been killed. Having worked several cases involving prostitutes, I've learned that many of these women lead mobile, nomadic lives. For one thing, they're often trying to stay one step ahead of the law. For another, they can command higher prices when they're the new girl on the block. So maybe the unaccounted-for prostitutes had simply moved on to greener pastures; on the other hand, maybe some of them were dead and decomposing in the woods or the old livestock barn. Unfortunately, the barn had gone up in flames during the summer and the site had been bulldozed clean. Was it an accident or arson? Any evidence that might have been there, including burned bones, was long gone. That left Cahaba Lane.

Six days after Patty Anderson's body was found, I got a call from the Knox County Sheriff's Department. They'd found the bodies of two more women out at Cahaba Lane, the officer said, and they wondered if I'd come take a look. I rounded up a team—Bill Grant (who later worked as a forensic anthropologist for the U.S. Army) and Lee Meadows and Murray Marks (both now UT professors who teach, work forensic cases, and keep the Body Farm going)—and we piled into a white UT pickup truck and headed east. A serial killer was on the loose in Knoxville, and he was preying on some of the city's most vulnerable women. Women whose livelihood required them to put their bodies—their lives—in the hands of strangers.

It had been years since I'd worked a serial-murder case, but I vividly remembered how disturbing it was. Back in the mid-1980s, eight women in the Southeast were murdered and dumped alongside major highways; three of the bodies were found in Tennessee. Many of the victims had reddish hair, so the case became known as the Redhead

Murders. Most of those women were prostitutes; that's when I learned how they'd move from one city to another whenever their earnings began to drop.

The Redhead Murders were never solved. I hoped this case would turn out better. There's no such thing as a happy ending in a case like this, but if we got lucky and all did our jobs well, at least there might be less crime and more punishment.

When I parked the truck at the end of Cahaba Lane and got out, I happened to glance down. There, clinging to the top of my left rear tire, was a slimy used condom. The investigators led us into the woods. The first body was about fifty yards to the right of the billboard—practically within sight of the pavement. This woman, like Patty Anderson, was partially clothed, although her pants were pulled down, exposing her buttocks and genitals. A black female, she was still in the first stage of decomposition: little discoloration, no bloating, minimal insect activity. That was partly because the body was fresh but also because the weather was cold. Blowflies don't fly if the temperature's below 50 degrees Fahrenheit.

"This body's too fresh for me," I said. "She needs to go to the medical examiner." Having recused myself from examining her, I was careful not to touch her. Just by looking at her bruised neck and contorted face, though, I was pretty sure she'd been strangled.

A sheriff's deputy asked how long she'd been dead. Just from a glance, and without making much allowance for the cold snap we'd been having, I said, "Not long—maybe a couple of days." That offhand remark, written down by the deputy and quoted in the newspapers, would come back to haunt me many times in the months and years to come.

They led me to the second body. This one lay much deeper in the woods than the others had, about a half-mile from the billboard, over the top of the hill and partway down the other side. Unlike the previous bodies, this one was completely nude; a satiny undergarment, a teddy, lay crumpled in the leaves about ten feet away. It was another black female, the race obvious from the hair and the exposed teeth. This body was badly decomposed. The skin was discolored and the abdomen bloated; the bones of the left leg were exposed; and both feet were missing. Legs and arms splayed wide, the corpse lay with its crotch jammed against a small tree. The tree trunk extending straight upward from the genitals of the nude, rotting body of the murder victim made the crime even more shocking, more depraved.

As I studied the body's position, I realized that this wasn't the death scene—in other words, this wasn't where she'd actually been killed. Looking around, I saw a dark, greasy stain several feet higher up the slope, where volatile fatty acids had leached out of the body. Part of the hair mat was there too. Clearly that was where her body had originally lain, until someone or something had come along and moved her.

Both of the victim's feet were gone, chewed off at the ends of the tibia and fibula, and the left thigh was badly gnawed as well. I could picture exactly what had happened: After the murder, a week or so passed; by then she would have been smelling pretty foul to you or me. To a dog's way of thinking, though, she was just starting to smell really interesting.

Dogs, I've observed, don't like to eat out in the open; they're afraid of being surprised from behind. Their favorite feeding position is to nestle with their back up against a log or a big rock, so nothing can sneak up on them. Now, if you're a 50- or 75-pound dog, and you're trying to drag a 120-pound body off to a safe place to eat it, you're not going to drag it uphill; you're going to grab a foot and drag downhill, so you get some help from gravity. In this case, though, the body hadn't moved far before the spread legs slid to opposite sides of a tree trunk. Once it was lodged there, the dog was stymied. Instead of a whole body, he had to settle for gnawing the thigh and carrying off the feet.

The body was lying faceup—except that the face was already gone. The soft tissue of the neck was also gone, exposing the cervical vertebrae, though the shoulders and arms remained virtually intact.

I wasn't surprised about the face; it's often the first thing to go.

Blowflies lay their eggs in moist, dark places, so the mouth, nose, eyes, and ears are among the obvious locations. So are the genitals and anus, if the flies can get to them. About the only place a female blowfly would rather lay her eggs than a natural body orifice is a bloody wound.

But while the missing face was to be expected, the missing neck wasn't, especially considering the good condition of the shoulders and arms. It was a classic case of what's called "differential decay," and anytime I see it I consider it a red flag, a clue. The differential decay in the neck region told me that there had been some sort of trauma there. Maybe her throat had been cut, in which case the flies would have flocked to the wound, or perhaps she'd been strangled, and her attacker's fingernails broke the skin and drew blood. Something, at any rate, had made the neck just as appealing to the blowflies and maggots as the moist orifices of the head.

As I studied the body Arthur Bohanan, a KPD crime lab specialist at the scene, spoke up: "Bill, give me a hand." Having worked with him for years, I knew he wasn't speaking figuratively. He wanted me to remove one of the victim's hands and give it to him.

Art was KPD's top fingerprint specialist; in fact, he was becoming known as one of the best fingerprint guys in the country, someone even the FBI consulted at times. He wasn't just a technician, dusting for prints at crime scenes; he was a researcher, exploring new ways to reveal latent prints on surfaces where they'd never been seen before, like fabrics and even the skin of murder victims. Art had worked a number of child abductions and murders over the years, and he'd seen children's fingerprints disappear—fade away from the interior of an abductor's car, for instance—far faster than the adult prints did. Why? Art decided to find out. Before puberty, he eventually learned, children's prints lack the oils that give adult prints such staying power.

To a civilian bystander, Art's casual request, "Give me a hand," would have sounded horrifying. To a forensic scientist it was routine. In a murder case it's not uncommon for investigators to cut off fingers or even entire hands to take back to their own labs or to send to the FBI. In any case when a victim's identity is unknown, it's important to try every possible technique to latch on to a print and a name. In a serial murder case like this one, the stakes were at their highest: At least three women were already dead, and if this killer fit the pattern of most serial killers, women would keep on dying until he was caught. This was no time for a squeamish sense of propriety.

I studied the hands. The skin was soggy and on the verge of sloughing off, but I knew that wouldn't keep Art from getting the prints: he'd been known to put his own fingers inside the sloughed-off skin of a murder victim's fingers in order to restore the natural contours and get the prints. From my point of view, the key question was whether the hands held any clues to the woman's manner of death or time since death. Examining them closely, I saw no defensive wounds, so she hadn't been fighting off a knife attack; there were no rope marks, no trauma of any sort.

Taking a knife from my tool bag, I cut off one hand, then the other, to double his chances of matching a print. I sealed them in a plastic bag and handed them to Art, who headed off to work his magic. On his way out he stopped to take prints from the fresh corpse down near the road, and he sealed those prints in another small bag.

For my work I would need a much bigger bag. On the ground beside the body, we zipped open a black "disaster bag"—the euphemism for *body bag*—and gently slid her inside the long opening. Then, with half a dozen of us gripping the corners and sides of the bag, we carried her out of the woods and put her in the truck.

As we were loading up, a police radio crackled to life. Art Bohanan had already ID'd one of the victims. Not the one whose hands he'd taken—that one would require more work—but the fresh one. Her name was Patricia Ann Johnson; thirty-one years old, she was a Chattanooga native who'd been staying in a Knoxville shelter for the past few weeks. She'd never been arrested for prostitution, but she'd been seen hanging out in the areas often worked by Knoxville hookers. Art relayed two other interesting pieces of information: She suffered from epilepsy, and her neck had several latent prints on it, which he'd detected by fuming her entire body with superglue and dusting it with ultraviolet powder. Unfortunately, there wasn't enough detail in those prints to identify the person whose hand had been squeezing her neck.

Now it was my turn to get to work and see what I could find out about victim number three.

We got back to the Body Farm just before dark. After I backed the truck through the gate, we pulled the bag out, laid it on the ground, and unzipped it to remove the body and begin cleaning off the tissue.

When we'd slid the body into the bag, we'd seen very few maggots—barely a handful. Now there was a huge swarm of maggots, literally tens of thousands of them. One of the students asked where they'd all come from. Could there have been a massive egg hatch during the forty-five-minute trip back to the university? No, I explained, just some confusion about what time of day it was. Maggots don't like sunlight, so if a body is out in the open, they burrow beneath the skin during daylight. When we sealed the remains in the opaque black bag, though, the maggots thought night had fallen, so they came out to feed on the surface.

One other interesting but gruesome note about maggots: Although cold weather keeps blowflies grounded, it doesn't faze their larval off-spring, the maggots. Even though we think of insects as "cold-blooded," as maggots digest human tissue, the chemical breakdown of the flesh generates a surprising amount of heat; on cold mornings at the Body Farm, it's not uncommon to see steam rising off a writhing mass of mag-gots huddled together for warmth. As my colleague Murray Marks has observed, for the residents of the Body Farm it's not quite as cold and lonely out there as you might think.

We attached metal tags to one arm and one leg to identify victim number three. This was our twenty-seventh forensic case in 1992; that meant she was case number 92-27. To estimate her age, we looked at several different bone structures: her skull sutures, her clavicles, and her pelvis. The bones of the pelvis were dense and smooth, with a marked absence of grain; in other words, they were the bones of a mature but young woman, probably somewhere between twenty and thirty. Her clavicles, on the other hand, had not fully matured: The medial, or sternal, end of the collarbone is the last piece of bone in the body to fuse completely to its shaft; the fact that this epiphysis, as it's called, had not yet fully ossified suggested that she was not yet twenty-five. Luckily, we could pin it down even more precisely than that. Research data from one of my former Kansas students indicated that the victim was probably somewhere between eighteen and twenty-three. Finally, the basilar suture in the skull—the joint where the occipital bone (the back of the head) meets the sphenoid (the skull's floor)—was only partly fused, another indicator that she was not yet twenty-five. Factoring all those indicators together, I was confident that she was somewhere between twenty and twenty-five.

To determine her stature, we measured the length of the left femur—44.4 centimeters—and plugged that value into a formula developed back in the 1950s but more recently refined a bit by a UT colleague, Dr. Richard Jantz. One of the world's leading authorities on skeletal measurements, Richard has assembled a huge database of skeletal measurements; he's also developed a powerful computer software package that, from a few simple skeletal measurements, can accurately determine an unknown corpse's sex, race, and stature. The stature calculation told us that with her 44.4-centimeter femur, our victim had stood about five feet three inches tall.

Now we knew sex, race, age, and stature. Next came the search for evidence of the manner of death. We checked and rechecked everything. There was no sign of trauma—no fractures, cut marks, or other traces of trauma—on any of the bones we had. But we didn't have every bone. Her feet were missing, but they probably wouldn't have told us how she died. One other bone was missing, though, and it was potentially the single most important bone in her body. It came from the region where differential decay had raised a red flag the instant I saw the body. What we were missing was the hyoid bone from the neck—the one bone that can reliably reveal whether someone has been strangled.

Floating above the larynx and below the mandible, or lower jaw, the hyoid is a thin, horseshoe-shaped bone. If you tilt your head back slightly, clasp the front of your windpipe, and wiggle your hand back and forth, you'll probably be able to feel your hyoid moving. From its exposed position and thin structure, you'll also understand why it's often broken in cases of strangulation.

Given that the two more recent victims had been strangled, it seemed crucial that we find the missing hyoid. We checked the body bag carefully, in case the hyoid was somewhere at the bottom of the bag, but no luck. I called together four of my grad students. "I need you to go back out to Cahaba Lane and find that hyoid," I told them. They looked dismayed and dubious, but I wasn't ready to give up. Time after time I've been amazed at how much skeletal evidence can be recovered from a death scene, even months or years after a murder: bones, bullets, teeth, even toenails. "Start where we found her," I told the students, "then work your way uphill to the spot where the hair mat was. It's got to be there." I meant that last part in more ways than one.

A few hours later they returned, triumphantly bearing the hyoid. Sure enough, up near the initial death scene, the bone had fallen out (or been plucked out by some scavenger), and then had been covered by falling leaves.

The hyoid was in three pieces, but that didn't necessarily mean it had been broken; in some people the hyoid never fully ossifies into a single arch of bone. Instead—as in this woman's case—the two side pieces, called the "greater horns," are joined by cartilage to the central arch, or "body." It was possible that the horns had been broken off, but it was also possible that the cartilage in those joints had simply decomposed. To know which was the case, I needed to look closer—much, much closer.

I took the pieces to a scanning-electron-microscope lab in the college of engineering. At a magnification of 20×, I thought I saw some traces of damage to the bone itself: tiny linear fractures and avulsion (literally, "pulling apart") fractures at the surface where the cartilage had been attached. I zoomed in for a closer look. Sure enough, at 100× and 200×, the damage was unmistakable: numerous microscopic linear fractures that ended in a small region of avulsed bone.

It wasn't much to look at, but it was crucial evidence: a telltale sign that cartilage had been ripped from this bone by some powerful force— for example, a pair of strong hands, squeezing mercilessly until the moment she ceased to struggle, ceased to breathe, ceased to live. That moment had probably come somewhere between ten and twenty days ago. I arrived at that estimate of time since death, or TSD, by correlating two sets of observations: the body's advanced state of decomposition, and the pattern of daytime and nighttime temperatures over the past several weeks.

To narrow down the TSD, I enlisted the help of my former student, chemistry whiz Arpad Vass, who was now a research scientist at Oak Ridge National Laboratory. I sent Arpad two soil samples: one taken from beneath the victim's body, where volatile fatty acids had soaked into the ground; the other an uncontaminated control sample taken from the hillside some fifteen feet above the death scene. In the Ramsburg case—the man shot by his wife, then buried in the crawl space beneath the house—Arpad had been handicapped by the lengthy postmortem interval. Here, though, the conditions were perfect for his technique. First, Arpad analyzed the relative concentrations of the decay products, then he factored in the temperature patterns. This time the technique worked brilliantly: Arpad's calculations put the TSD at fourteen to seventeen days. Based on the decomp, I had put the murder sometime in the period of October 6 to 16; Arpad narrowed that window to October 12 to 15, right around the same time Patty Anderson had disappeared.

Just to be sure, the sheriff's investigators had asked for an additional TSD estimate, for each of the bodies, from a forensic entomologist named Neal Haskell, who had done an interesting study at the Body Farm a few years before. Neal was developing a forensic technique for re-creating death scenes, in effect, by using a freshly killed pig as a stand-in for the murder victim—a "body double," as they say in Holly-wood, but of a different species. By letting nature take its course until the insects on the pig carcass matched those on the human victim, Neal hoped to be able to pinpoint time since death to within a day or two. But to know whether the pig carcasses would work as stand-ins for hu-mans, he needed to make a head-to-head comparison of the bug activ-ity in each species. The only place where he could do that, of course, was at UT's Anthropology Research Facility. I was glad to have him do the study there; if the technique worked—and the study showed that it did, at least within the first couple of weeks postmortem—it could be useful at crime scenes virtually anywhere.

When he was called about the Cahaba Lane murders, Neal immedi-ately set about getting samples of live maggots from the bodies so that he could time how long it took them to mature into adult flies. It's an entomologist's way of figuring out when the eggs were laid, like count-ing backward from a baby's birth to figure out when it was conceived.

Neal also put several pig carcasses in the woods at Cahaba Lane; sheriff's deputies were posted to guard the experiment and to take tem-perature readings at frequent intervals. Judging by how long it took the maggots from the bodies to mature, together with what he observed in the pig carcasses, he calculated that blowflies had first begun laying eggs on this woman's body sometime between October 9 and 13. So three different scientists, using three different techniques, agreed pretty damned closely on when she was killed.

My final challenge would be to find out who she was. With luck, I could learn that straight from her very own mouth. Her teeth were a study in contrasts. On the one hand, a lot of careful work had gone into that mouth: fourteen of her teeth had amalgam fillings. On the other hand, one of her teeth, the lower left first molar, was literally rotting away. The cavity had eaten away much of the crown and spread down into the tooth's pulp chamber; as a result, the jawbone itself was begin-ning to crumble too.

I'd seen this sort of contrast before, especially in females. Almost invariably, it pointed to a dramatic change in the victim's fortunes. A girl grows up, leaves home, and has a hard time making her way in the world; an older woman gets laid off, divorced, or widowed. Whatever the cause of the setback, she cuts costs and corners wherever she can, and before long, dental care is a luxury she can no longer afford.

But even though 92-27 had fallen on hard times, somewhere out there—from the period before her life started going wrong—there were dental X rays with her name on them. I knew we could find them, but it might take a while. Fortunately, we were spared the trouble.

While my colleagues and I had been scrutinizing teeth and bones, chemicals and insects, KPD fingerprint wizard Art Bohanan had been working with the hands I'd cut off for him at the scene. The police had no prints on file that matched the ones Art lifted from the hands, so if she'd ever been arrested, it was someplace besides Knoxville. She also didn't match any police descriptions or profiles of known prostitutes. But her general description—black female, age twenty to twenty-five, height five feet three inches—did match a missing-person report filed recently by a local woman's sister. The missing woman, last seen on October 14, was Darlene Smith, a black female age twenty-two, height five feet four inches—a mighty close resemblance to the woman the skeletal analysis described.

From her sister's report, Art had Darlene Smith's address, a rented apartment in the eastern section of Knoxville, not far from an area frequently worked by prostitutes. The neighborhood wasn't particularly desirable but it was pretty cheap. The sister let Art into Darlene's apartment and dug out a copy of her lease. Art sprayed the paper with ninhydrin, a chemical that reacts strongly with the amino acids in human fingerprint oils. Within moments a jumble of bright purple smudges and prints appeared before his eyes.

The prints came from two pairs of hands, Art determined. One pair of those hands belonged to a man—Darlene's landlord, Art learned by fingerprinting him that night. The other prints on Darlene Smith's lease matched the hands I had severed from the decaying corpse at Cahaba Lane.

THE MORNING OF OCTOBER 27, the phone rang again. The police had just found a fourth victim in the woods. I nabbed Bill Grant and Lee Meadows, who'd gone with me the day before, and Emily Craig, the Ph.D. student who had taught me the difference between Caucasoid knees and Negroid knees. Together we retraced the now-familiar route to the scene.

The fourth body lay about a quarter-mile to the right of the bill-board, at the edge of the small creek emerging from the woods. Wide and flat, the streambed was dry for much of the year; now, though, a few inches of water trickled through it.

The body was largely skeletonized, except for areas of tissue on the legs, buttocks, and left arm and hand. Lying faceup amid the oak leaves, the bare skull fixed us with a sightless, accusing stare. The vertebrae were completely defleshed, covered only by leaves and twigs. The right arm and hand were missing, probably chewed off by a dog. The left hand, though, lay in the streambed, covered with mud and water. As I dug around it carefully with a trowel, I was pleasantly surprised to find that some of the hand's soft tissue was still intact.

We bagged the remains and took them back to UT Medical Center. Our first stop was the hospital's loading dock, where we used a portable X-ray machine to check for bullets, a blade, or any other foreign objects that might tell us something. But there was nothing metallic in the skeleton of this victim, 92-28, except for some dental fillings. Next stop was the Body Farm, where we set the corpse on the ground, opened the body bag, and began cleaning the remains.

Art Bohanan had followed us back from Cahaba Lane. I knew what he wanted, but he wouldn't have much to work with this time. Not only was there just one hand, there wasn't even a whole lot of that one. The entire thumb was gone; so were half of the index and middle fingers. About all that remained were the ring finger, the little finger, and part of the palm. But if anybody could tease out an identifying print from a fragment of a rotting hand, it was Art.

Because the remains were already virtually skeletonized, it took me less time than usual to clean the bones for a forensic examination. I could already tell, even out in the field, that this was a woman. The pelvis was textbook female: broader hips, a raised sacroiliac joint, a wide sciatic notch, and a greater subpubic angle—all part of the geometry designed to allow a baby's head to pass through the pelvis during birth. The cranium, too, had classic female features. The upper edges of the eye orbits were sharp, the chin tapered to a point at the midline, and the cranial vault was smooth and lacking in heavy muscle markings.

Race was easy to peg, too. On the ground beside the skull we'd found the hair mat where it had sloughed off: light brown and slightly wavy. That hair, plus the shape of the mouth—teeth that were oriented quite vertically, with no forward protruding or jutting—clearly marked her as white.

To estimate age, we looked at several different bone structures: her upper jaw, her clavicles, and her pelvis. Like 92-27, 92-28 had pelvic bones that were dense and smooth, with a marked absence of grain; in other words, they were the bones of a mature but young woman, probably somewhere between her mid-twenties and mid-thirties. Her clavicles were fully mature as well: the medial, or sternal, ends of the bones had fused completely to the shafts, which meant she was at least twenty-five. Finally, her cranial sutures—including those in the hard palate, called the intermaxillary sutures—were not yet fully fused. Generally the intermaxillary sutures don't fuse until the late thirties, so she probably wasn't more than thirty-five. I could say with certainty, then, that she was somewhere between twenty-five and thirty-five, but it was hard to be more precise than that.

You'd think, since the skeleton was missing only an arm, that we could determine her stature simply by laying the remains on a lab table and stretching a tape measure from head to heel. Unfortunately, it's not quite that simple. After death, cartilage shrinks and decays, sometimes by as much as several inches. Besides, her skull wasn't attached. Those two complications virtually guaranteed that the tape-measure method would be wildly inaccurate.

Instead, we used the same method we'd have used if we'd found nothing but a single femur: we measured its length and extrapolated. This femur was longer than the previous one: 47.8 centimeters. That put 92-28 at somewhere between five feet six and a half inches and five feet nine and a half inches tall.

Next, I sought any signs of trauma that might tell me how she died. Unfortunately, despite hours of sifting through the leaves and soil, we'd never managed to find her hyoid, so I couldn't tell if she'd been strangled.

Another bone, though, did reveal something striking—literally. The left scapula, or shoulder blade, had a large fracture on its lower end. Now, the scapula's a pretty big, strong bone, and it's well protected by large muscles. That fracture could only have been caused by a powerful blow—maybe a violent kick from a heavy boot, or possibly a hard hit by a baseball bat or a two-by-four.

The pattern of breakage at the edges of the fracture indicated that the blow had come from behind, and there were no signs of healing, so the fracture was perimortem (occurring at or just before death). In other words, she was probably running for her life when he caught up with her. She was barefoot, remember, and he was surely wearing shoes. He knocked her facedown on the stream bank, then he set upon her and killed her.

The greater the length of time since death, the harder it is to pinpoint, at least from the skeletal remains. Because the corpse was almost fully skeletonized, it was obvious that although 92-28 was the last to be found, she'd been the first to die. Taking into account the extreme decomposition of the body, the daily temperatures in September and October, and the condition of the soft tissue that had been submerged in the creek—whose decay rate I knew would therefore have been cut in half—I judged that 92-28 had been dead four to eight weeks before she was found, a pretty wide window of opportunity covering almost the entire month of September. The bugs and the soil analysis, I hoped, would tell the time of the crime with considerably more precision.

My hopes proved well founded. Arpad's analysis of the volatile fatty acids from the soil beneath the body put the TSD at thirty to thirty-seven days, meaning that she had been killed sometime during the week of September 22 to 29. Neal Haskell's entomological analysis came to virtually the same conclusion: September 22 to 26. If indeed she'd died during late September—the interval where our three independent analyses, based on different techniques, overlapped—then the time between killings fit the classic, accelerating pattern for serial killers: Two to three weeks had elapsed between the first murder and the second; perhaps a few days between the second and the third; and, according to the autopsy findings of the medical examiner (ME), as little as a day or two between the third and the fourth.

This victim's teeth, like Darlene Smith's, showed a pattern of careful attention during her youth, followed by neglect and decay in recent years—in other words, another mouth that had fallen on hard times. Half a dozen teeth bore fillings, but one tooth, a lower left incisor, had two unfilled cavities. One of these was small, but the other extended from the top surface of the tooth deep into the pulp cavity. This cavity had probably been filled at one time, but the filling had fallen out, making the tooth more vulnerable to decay than ever. The infection had spread to the jaw itself, causing a large abscess on the surface of the bone. When I first picked up the skull out at the crime scene, I'd noticed that this hole in the tooth had been stuffed with cotton. At the time I said to Art Bohanan, "She had a toothache when she died"; I thought the cotton indicated that a dentist was about to perform a root canal. As it turned out, the police later learned that she was self-medicating with a unique and desperate remedy for the pain: before inserting the cotton, she was soaking it with a paste of cocaine. Desperate times, desperate measures.

Once more, just as he had with Darlene Smith, Art Bohanan took the hand he was dealt and hit the jackpot. What little skin remained on the hand was waterlogged, decomposing, and incredibly fragile. Art soaked it in alcohol to toughen it up and draw out the water. (If he'd had the opposite problem—if the skin had been dry and stiff—he'd have soaked it in Downy fabric softener; I'm sure the makers of Downy would be pleased to know that their product makes even mummified human skin soft and fragrant.) He'd only been able to salvage one print from the ravaged hand, and it wasn't even a fingerprint. All he could get was a partial palm print, from the edge of the palm just below the pinky finger.

It wasn't much, but it was enough. That partial palm print matched a print on file at KPD: It belonged to Susan Stone, age thirty, height five feet nine inches. A prostitute and a cocaine addict, she'd started her downhill slide some seven years before, when she married a drug dealer. She'd worked some conventional jobs before becoming a hooker; in fact, just six months before she died, she was working as a clerk at a data-processing company. If she'd hung on to that job, she might have hung on to her life.

CATCHING A SERIAL KILLER is a mammoth job, requiring teamwork every step of the way. Identifying the murder victims, determining how and when they were killed, and following the trail of evidence to the Zoo Man's door required the combined efforts of police investigators, a forensic pathologist, forensic anthropologists, a research scientist, and a forensic entomologist. This case is the best illustration I know of such teamwork. Bringing a serial killer to justice is an equally mammoth job, one that can drag on long past the point when someone is arrested and charged with the murders. This case is the best illustration I know of that, too. As my colleagues and I had labored to coax whatever evidence we could from the bodies of the murdered women, the police struggled to coax evidence from Tom Huskey.

Two weeks after his arrest, their efforts finally began to pay off, and in spectacular fashion. In a series of interviews, Huskey confessed to murdering the four women. As a tape recorder captured the lurid details, he told detectives how he shoved one body (Patty Anderson's) under a mattress and took her necklace and earrings—items police found in his room when they arrested him. Huskey described his final victim as a black woman who was tall, thin, and "ugly." She got scared, he said, and started having a seizure of some sort, flopping around "all over" the ground. His report was consistent with the physical description and medical history of Patricia Johnson, the recent transplant from Chattanooga whose body I had pronounced too fresh for me to examine.

It wasn't long, though, before the tape-recorded sessions took a bizarre turn. When the taping began, Thomas Huskey was speaking softly, almost meekly. Soon, though, his voice changed dramatically: It grew loud, belligerent, profane—and it belonged to someone else, another personality it called "Kyle," Huskey's evil alter ego. "Kyle" bragged that it was he, not Thomas, who had committed the murders. Then came a third voice, cultured, with a British accent. This voice said to be "Phillip Daxx," an Englishman born in South Africa, who said his role in the trio of personalities was to protect Tom from the evil Kyle. On one level, the case against Huskey appeared ironclad. But the bizarre claims of the various voices complicated the picture enormously. And Huskey had another powerful factor working on his behalf: the toughest defense lawyer I've ever seen. Herb Moncier was legendary in Tennessee for his aggressive tactics, his willingness to fight tooth and nail for his clients.

Moncier wasted no time before going on the offensive. Filing motion after motion, he sought to have Huskey's confession thrown out; he sought a new venue, arguing that all the newspaper and television coverage made it impossible for Huskey to secure a fair trial in Knoxville; he moved to have Huskey declared mentally incompetent to stand trial; he demanded that the judge recuse himself from the case; he demanded more time, more psychiatric evaluations, more money for the defense.

Under the barrage of defense motions, the murder case ground to a halt. But I was too preoccupied to notice or care whether a jury sentenced Zoo Man Huskey to live or die. I was preoccupied with a far more urgent life-or-death struggle.

For decades I had worked closely with mortality. It was almost as if I donned some charmed cloak of immunity every time I strode cheerfully into the valley of the shadow of death. We had an arrangement, the Reaper and I: I would follow in his footsteps, and he would leave me alone. Our relationship was close but strictly professional. Then one day it turned personal. Unfortunately, it wasn't me he was after. He reached for the person who had walked by my side for forty years.

IN THE FALL OF 1951, the Korean War battles of Bloody Ridge and Heartbreak Ridge loomed large in the minds of most young American men, including me. Fresh out of the University of Virginia, I was due to be drafted for military service. On November 15, I reported as ordered to the armed forces induction station in Martinsburg, West Virginia. I was one of about two hundred draftees processed that day. The sergeant handling our induction called out the first fifteen names on his list—it was alphabetical, so I was number two or three—and assigned us to the Marines. My heart sank. The Marines were taking the brunt of the U.S. casualties in Korea, so I thought I was lost.

Just then a lieutenant intervened. The lieutenant noticed on my intake papers that I'd graduated from UVa and had taught math and science. Figuring that I might be reasonably bright (or possibly not seeing me as one of the Few Good Men the Marines were looking for), he told the sergeant to assign me to the U.S. Army instead, in the "scientific and professional" category. The sergeant objected; the lieutenant persisted. When the sergeant continued to argue—in front of a roomful of draftees—the lieutenant finally pulled rank, snapping, "That is an order, Sergeant."

I was saved. Instead of the Korean Peninsula, I was sent to the Army Medical Research Lab, called AMRL, at Fort Knox, Kentucky, to help study how noise and vibration—from trucks, tanks, and artillery—affected the soldiers using them. I would spend the rest of the war surrounded by dozens of doctors, research scientists, good-looking nurses, and deafening, powerful machines. Life was good. Then it got even better: I met Lieutenant Owen.

An old friend of my mother's was stationed at the Pentagon outside Washington, D.C.: Colonel Hilda Lovett, the senior dietician for the Army's entire network of hospitals. Colonel Lovett had promised my mother that she'd look out for me, and she was true to her word. When she heard I was assigned to AMRL, she cast her eye about for a suitable girlfriend for me, and her eye happened to light on a bright young nutritionist in training at Walter Reed Army Hospital: First Lieutenant Mary Anna Owen. Lieutenant Owen was scheduled for assignment to Fort Lee, in Virginia; however, whether through coincidence or through meddling at the highest levels of the Pentagon, her orders changed and she went to Fort Knox. I received orders of my own: I was to call on this lieutenant and make her feel welcome.

On the appointed afternoon in the fall of 1952, I arrived at her apartment. As always, I was compulsively early, but when I got there she wasn't in; she was next door, chatting with another nutritionist. Hearing me knocking, she came running. When I heard footsteps and turned, what I saw wasn't Lieutenant Owen double-timing in an Army uniform; what I saw was a girl named Ann, glowing in a red dress. The instant I saw her running toward me in that red dress, I thought, *That's the girl I'm going to marry.*

And I was right. Less than a year later we got married in my hometown in Virginia, in the presence of my mother, my stepfather, a horde of friends and relatives, and the person who had made the match, Colonel Hilda Lovett.

Ann and I spent the next forty years building a life together. Between the two of us, we earned four graduate degrees and produced three healthy sons. Life wasn't always easy; between our first child, Charlie, and our second, Billy, Ann suffered five miscarriages. But on the whole, we were blessed, busy, and happy.

We moved from Fort Knox to Lexington to Philadelphia to Nebraska to Kansas to Tennessee. We spent a dozen summers in South Dakota, where I spent my days digging dead Arikara Indians out of the ground and Ann spent hers keeping live Sioux out of the ground, helping the tribe fight diabetes through better nutrition. Before we knew it our sons were grown, and in August of 1990 our first grandchild arrived. A new chapter in our lives was beginning. But it didn't end the way we expected or wanted. One year later, Ann got sick.

It began with abdominal pain—intermittent at first, then constant. Ann went to our family doctor, who took a stomach X ray. The radiologist noticed what looked like an obstruction at the very edge of the film, in the lower GI tract, so Ann went to a hospital, drank that awful barium milkshake, and had a fluoroscopic exam. The pathologist told us it was cancer, and it was pretty advanced: well into Stage Three already, which meant it was probably spreading elsewhere in her body.

Ann wanted to fight it. At sixty, she was still a relatively young woman, and she was looking forward to lots more grandchildren, so she embarked on a course of aggressive chemotherapy. The chemo took a heavy toll on her, but she endured the treatment until it was too late. In March of 1993, eighteen brutal months after that first visit to the doctor, Ann died.

For decades I'd dealt with death on a daily basis, but I'd always managed to float untouched by the tragedy surrounding me. I was a scientist; to me, decaying bodies and broken bones—my stock and trade—were forensic cases, scientific puzzles, intellectual challenges—nothing more. That's not to say my heart was hard, that it didn't go out to the people whose loved ones had died; it did, especially to the parents of murdered children. But those were passing waves of sympathy. Now that death had finally hit home, I was drowning in an ocean of grief.

THE ZOO MAN CASE dragged on, throughout Ann's illness and beyond, with no murder trial anywhere in sight. Meanwhile other women had come forward to say that Huskey had assaulted them. In late 1995 and 1996, Huskey stood trial for a series of brutal rapes in 1991 and 1992.

Moncier lost that case, one of his few high-profile losses I can recall. Huskey was found guilty on various counts of rape, robbery, and kidnapping and was sentenced to sixty-six years in prison for three rapes and a robbery. But the murder case remained stalled by Moncier's barrage of motions and maneuvers. Finally, in January of 1999—more than six years after the four women were killed in the woods off Cahaba Lane—jury selection began for Huskey's murder trial. Moncier hadn't managed to get the trial moved; however, he did prevail on the court to import jurors from out of town, in hopes they would be less likely to have been swayed by Knoxville's extensive news coverage of the case.

An initial pool of 340 potential jurors was called, then narrowed down to 60. Some prospective jurors were desperate to be released from jury duty; others were equally eager to serve. District Attorney Randy Nichols had indicated he would seek the death penalty, so jurors who said they were unequivocally opposed to the death penalty were excused. After a couple weeks of interviews in Nashville by the prosecution and the defense, twelve jurors and four alternates were told to pack their bags, then bused to Knoxville. For the next two weeks they would spend their days in the courtroom and their nights in an undisclosed hotel.

On January 26, 1999, the Zoo Man murder trial finally got under way. The linchpin of the prosecution's case was Huskey's own confession, in which he described the murders in detail. But if the confession made it clear that Huskey—or "Kyle," or whoever he called himself that day—had strangled the four women, the tape gave the defense some powerful ammunition as well. The trio of voices and names played back through the speakers made it easy to believe that Zoo Man really was crazy. To buttress the insanity defense, Moncier called witnesses ranging from a psychiatrist and a psychologist—both of whom agreed that Huskey suffered from multiple-personality disorder—to Knox County jail workers who testified that they'd talked with Huskey's evil alter ego, "Kyle." Curiously, Huskey's mother denied any knowledge of "Kyle" or "Daxx." Tom, she said, was just plain Tom: That's all; there was nobody else in there.

The defense didn't challenge my analysis of the scapula fracture. The hyoid, though, was another matter entirely. The electron micrographs clearly showed trauma to the bone, but Moncier disputed my conclusion that it indicated strangulation. He called his own expert witness, a pathologist from Atlanta—who was a physician, true, but who wasn't board-certified. The pathologist ventured that maybe a deer had stepped on the hyoid and crunched it; Moncier pressed me as to whether that was possible. Well, hell, anything's possible. It was possible a Martian spaceship had landed on it, but the only explanation that satisfied both forensic science and common sense was that the woman had been strangled.

The trial itself lasted for two weeks, then the jury began to deliberate. The deliberations dragged on for a day, two days, three. Eventually the jury sent out a note saying that they agreed that Huskey had killed three of the four women. As to the fourth murder, eleven of the twelve jurors were convinced of his guilt, but the twelfth juror thought it possible that the final murder had occurred after Huskey's arrest on October 22. (Although Neal Haskell's entomological analysis had put the murder around October 21 or 22, Moncier had hammered away at my offhand remark that Patricia Johnson might have been dead only "a couple of days.") Despite arguments and pressure from the other eleven jurors, the twelfth continued to hold out.

But in the end, the real stumbling block wasn't Huskey's guilt or innocence; the real stumbling block was his sanity. By the fourth day of deliberations the twelve jurors had divided into three immovable groups: Five believed that Huskey was sane and should be held accountable for the murders; four believed he was insane; the other three couldn't make up their minds. Finally, on the fifth day, they sent the judge a note saying they were hopelessly deadlocked.

After six years, half a million dollars, and thousands of hours of investigative work and legal wrangling, Judge Richard Baumgardner declared a mistrial. For the police, prosecutors, and victims' families, it was a bitter blow. But worse was still to come. In July of 2002, Judge Baumgardner—ruling on yet another defense motion—agreed to bar the use of Huskey's confessions. Twice during his interrogation—the day of his arrest and again a week later—Huskey had asked for a lawyer, but investigators with the Knox County Sheriff's Department and the Tennessee Bureau of Investigation continued to question him.

As of this writing, Tom Huskey's retrial on the four murder charges has been postponed—again—and an appeals court has reversed some of his earlier rape and kidnapping convictions and lowered his sentence to forty-four years. Legal insiders say the murder cases might be dropped altogether, if the confessions can't be used as evidence. The wheels of justice turn slowly, it seems . . . and sometimes they stop altogether or even spin backward. On the other hand, the man who confessed to killing four women remains behind bars, for the moment at least, and is scheduled to remain there another forty years. And the only bodies to emerge from the woods at the end of Cahaba Lane, during these ten years Huskey has been behind bars, have been a few bushy-tailed squirrels. Out on Magnolia Avenue, though, a new generation of women is working the streets again. Turnover's high out there. I wonder how many of them have even heard of Zoo Man. I wonder if they realize how vulnerable they are. I wonder, even if they do, if they can do a damn thing about it.

CHAPTER 13

Parts Unknown

THE PHONE RANG, startlingly loud in the silence. It was July, and the university was practically a ghost town. The hallways were dim and deserted in the depths beneath Neyland Stadium. Most of the students and faculty had vanished in late May and wouldn't reappear until late August. Understandably, they seized any opportunity to get out of the depths of the stadium. I, on the other hand, was spending nearly every waking moment down in my dark, dusty office. It had been months since Ann died, but I still couldn't bear the emptiness of our house. At work, by contrast, I was surrounded by people. Most of them were dead, mind you, but they were comforting all the same. They had shared their stories with me and had entered my life; they were companions who would never abandon me. Besides, at work, I knew, it wouldn't be long before someone would call me with an interesting case. So when the phone rang on this quiet summer day, I reached for it eagerly.

At the other end of the line was my secretary, Donna, whose office lay, literally, a football field away from my private sanctuary, tucked deep beneath the stadium's east stands. She was transferring a call, she said, from Corporal James J. Kelleher of the New Hampshire State Police.

"Hello, this is Dr. Bass," I said. Corporal Kelleher introduced himself. He worked in the major crimes unit, he explained, and was the lead investigator on a case he believed might involve a homicide. He had read about me in *Bones*, a book written by Doug Ubelaker, a former student, who was now a staff anthropologist at the Smithsonian Institution. (One of the things that thrills me when I look back on my career is the fact that three of the Smithsonian's physical anthropologists—Ubelaker, Doug Owsley, and Dave Hunt—got their Ph.D.s from me, and I served on the doctoral committee for a fourth, Don Ortner.)

As Kelleher outlined the case, I began to take notes. A few handfuls of burned bone fragments had been found in a yard in Alexandria, he said, a tiny hamlet in the center of the state. The medical examiner thought they were dog bones, but Kelleher suspected they were human. If he was right—if the bones were indeed human—he needed a positive identification of the dead person; if possible, he also needed to know the manner of death. Kelleher asked whether I could help. "I believe I can," I said. "I can sure try."

Six days later, a well-wrapped FedEx package arrived; inside the layers of paper and bubble wrap was a box containing bone fragments— hundreds of them—burned to a crisp. By this time I'd examined dozens of burned bodies and thousands of burned bones; they'd been sifted and plucked from burned-up cars, burned-down houses, even a "blowed-up" fireworks factory, as some of the locals would say. But except for bones from commercial crematoriums, I had never seen bones so completely burned as these.

Nearly every forensic case represents a scientific jigsaw puzzle, figuratively speaking. This one was a puzzle in the most literal way you can imagine. All told, the package contained 475 individual bone fragments, many of them no bigger than a pea. Piecing together even an approximation of a partial human skeleton would take days of tedious puzzle work.

I took the package down to the bone lab, in the basement of the stadium, where there was plenty of work space, good light from a wall of windows, and a stout lock on the door to protect the chain of custody. Clearing off one of the long tables near the windows, I unrolled a long piece of brown wrapping paper and taped it down. With a felt-tip marker, I wrote the names of the main sections of the body—skull, arms, ribs, vertebrae, pelvis, and legs—in their normal anatomical positions, more or less. Sorting the pieces into related piles would make it easier to start piecing together the charred rubble that had once been a human being.

Over the next few days I worked to reconstruct the life-size puzzle. The work was demanding, tedious, and baffling: exactly the kind of scientific challenge I've always liked best. Some pieces were fairly easy. There were four fragments from the right femur; remnants of both kneecaps; dozens of pieces of ribs; and three partial vertebrae. But all too soon I'd pulled out and placed every one of the big, easy pieces; all that remained were tiny, difficult pieces, and hundreds of them. A *challenge*, I reminded myself. *You always say you like a challenge. Be careful what you wish for.*

The pieces seemed to come from every major area of the body—all but one, I gradually realized: out of the 475 fragments, I couldn't find a single piece from the skull. That's not to say there wasn't one; more than half the fragments were so small and featureless that I couldn't tell what bone they came from. Still, the empty space at the top of my brown-paper chart seemed more than random coincidence. Worse, it meant I wouldn't be able to shed much light on who this was and how he or she had died.

Ten head-scratching days later, FedEx brought me another package from Jim Kelleher, smaller than the first but equally well wrapped. This one contained a large, relatively unburned piece of bone, easily recognizable as the mid-shaft of a left human femur; a glass vial containing more than five dozen small bone fragments; and one other bone, unburned but covered with tooth marks. Dogs, probably, had chewed off the upper end; the lower end had been broken off. Unlike all the other fragments, this bone was clearly nonhuman. I headed down the hall to consult one of my colleagues, a zoological archaeologist named Walter Klippel. Walter instantly recognized it as a tibia from the left hind leg of a white-tailed deer.

According to Kelleher, the first batch of burned fragments had been found on July 2, in a household pit used to burn brush and trash; the second set was discovered on July 22, scattered alongside a trail leading into the woods behind the house.

Unfortunately, I still didn't have a skull or teeth to work with; that meant I would probably not be able to make a positive identification from these remains. With a bit of good luck, there might be a healed fracture or some other distinctive feature on the bones that could be matched to someone's antemortem X rays. In this case, though, good luck seemed not to be in the cards.

Still, there was enough detail in the bones—burned and fragmented though they were—to help narrow things down quite a bit for Kelleher. One relatively intact piece of bone was the unburned head of the humerus, the ball at the end of the upper arm that joins the shoulder. With a pair of sliding calipers, I carefully measured its diameter at the thickest part. Back in the 1970s, T. Dale Stewart—a Smithsonian anthropologist whose close collaboration with the FBI in the 1950s and '60s helped pioneer the field of forensic anthropology—had made a careful study of humeral head size in males and females. According to Stewart's research, if the head measures more than 47 millimeters in diameter, it has to have come from an adult male. Measurements in the range of 44 to 46 millimeters can indicate either a male or female. A measurement of less than 43 millimeters unequivocally indicates a female. The piece lying on my lab table measured 42 millimeters; that meant our mystery victim was a woman, a finding borne out by a characteristically female ridge in the hipbone.

How old had she been when she died? Estimating age is easy if you have the pubic symphysis. Unfortunately—more bad luck—I didn't. Instead, I had to rely on several less-precise age markers. Judging by the fact that the ends of all her bones (the epiphyses) had fused to their shafts, I could tell that her growth had stopped. Okay, now I knew she was a grown woman. But she wasn't an old woman, because her spine showed only minor traces of osteoarthritic lipping, the jagged edges that vertebrae begin to acquire when we're in our late thirties or early forties. One other bone, the coccyx, or tailbone, showed surface features consistent with an age range of thirty-five to forty-five.

But that was about all I could tell Kelleher for sure. I couldn't even say whether she was Caucasoid, Negroid, or Mongoloid.

"I wish we had a skull," I told him.

FIFTEEN MONTHS LATER I got my wish. On a cold October night in 1994, I stepped off a Delta flight onto the windy tarmac in Manchester, New Hampshire. Kelleher met me in the terminal, helped me collect my suitcase, then dropped me at a hotel in Concord, the state capital. The next morning he picked me up and took me to the crime lab in the basement of the New Hampshire State Police headquarters.

Basements: Why are crime labs and morgues always in basements? Why not up on the top floor, with big corner windows looking out across the city or the countryside? Just because some of us like to look at bodies and bones, that doesn't mean we wouldn't appreciate a nice view out a window every now and then. But I'm getting sidetracked.

A bit of good luck had finally come our way. A road crew clearing brush along a cul-de-sac in Alexandria a few days before had stumbled upon a plastic garbage bag tossed into the weeds. Inside was a human skull, along with a number of other bones. Some, including the skull, were slightly burned; others showed no signs of burning at all.

A comparison of the teeth with dental X rays confirmed what Jim Kelleher had suspected for quite some time: the dead woman was Sheilah Anderson, a white female, age forty-seven, reported missing sixteen months earlier. Unable to contact her, Mrs. Anderson's adult daughter had phoned the police in June of 1993, about two weeks before the first batch of burned bones was found; that's why Kelleher had asked me to double-check the ME's initial impression that the burned bones were from a dog. Sheilah's husband, Jim Anderson, a former New York City police officer who had left the force under suspicious circumstances, told investigators that his wife had simply skipped out one day. She left, according to Anderson, for parts unknown.

Sheilah's daughter had doubted her stepfather's story. So had the state police, particularly after Jim Anderson attempted suicide in the days after his wife's disappearance. He was put in a hospital psychiatric ward for observation. On July 2, the day he was scheduled to be released, a trooper had accompanied Sheilah's daughter to the house so she could get some clean clothes for Jim to wear home. While she was at the house, she decided to take a look around. Out back, at the edge of the woods, she found a burned tennis shoe that she recognized as her mother's.

The trooper then began looking around in earnest. In the front yard he'd noticed ashes from a brush pile Jim Anderson had burned a few weeks before. Sifting through the ashes, he began to find bone fragments—the 475 charred pieces that were the start of my skeletal jigsaw puzzle. At that exact moment, Jim Anderson arrived home from the psychiatric ward. When he saw the officer pulling bone fragments from the ashes, Jim began drinking hard and fast. Vodka, straight up.

Ten days later, the police found the second batch of bones—the shaft of the femur, the deer tibia, and the vial of additional fragments—scattered in the woods near the burned sneaker. Then came the long wait—fifteen months—before the skull turned up. With the skull finally in hand, Kelleher no longer needed me to make a positive identification; the dental X rays had done that within hours after the highway crew found the garbage bag in the weeds. (A necklace of Sheilah's was still fastened around the vertebrae, as if to erase any speck of doubt.)

The mission that had brought me a thousand miles, to the basement of the New Hampshire State Police, was to shed whatever light I could on Sheilah Anderson's manner of death. One look at the skull and I knew the trip was not going to be wasted. The back of the skull was burned, but not much. Halfway up, slightly to the right of the midline, was a round hole the size of a silver dollar. I'd seen holes like this many times before: they're what's left by a blow from a hammer swung with great force against the cranium. The blow not only punched out a disk of bone, it also sent fractures racing outward like lightning from the point of impact.

On the inside of the blackened skull, surrounding the hole, was a dark, irregular stain: blood that had flowed from the wound, then cooked in the fire. The blood ruled out any possibility that the trauma to the skull had occurred when the garbage bag was chucked into the weeds. Postmortem wounds don't bleed once the blood has cooled and rigor mortis has set in. Sheilah Anderson was killed and then she was cooked.

The face of the skull wasn't burned, but it was broken: three of the upper front teeth had snapped off; the tips of both nasal bones were fractured; and the lower jaw was broken in three places. It was just the sort of trauma I'd expect to see in the face of a woman who got hit from behind with a hammer, then fell face-first onto a basement floor or a driveway.

What I *didn't* expect to see was the trauma in the other bones recovered from the roadside garbage bag. The fifth, sixth, and seventh cervical vertebrae showed cut marks from a large, sharp implement of some type. When I put the cervical (neck) and thoracic (chest) vertebrae together, aligned as they had been when she was alive, the damage was startling: an entire section of the spine had been cut loose from the ribs. The ribs on the right side had been severed close to the vertebrae; the left ribs had been cut farther from the spine, leaving stumps about two

My father holds me in our driveway in Staunton, Virginia, about 1930. One of my few memories of him is the trip we made into town in this car to buy a Sunday newspaper.
(Courtesy Jennie H. Bass)

Seventy years later, I take a break in my office beneath the stands of Neyland Stadium at the University of Tennessee. Beside me is a mounted display of human bones, which I use in teaching anthropology classes and forensic seminars.
(Copyright Ed Richardson)

An aerial view of the Larson, South Dakota, site, at the edge of Oahe Lake in 1970. The parallel strips mark scraper cuts, each about 120 feet long. The dark, round circles within the cuts are excavated graves. We started excavating at the right-hand side, working our way uphill just ahead of the rising waters. A year after this was taken, the entire site was under water. *(Collection of Dr. Bill Bass)*

(LEFT) Pat Willey spent many summers excavating Indian graves with me in South Dakota, starting when he was a junior-high student in my Sunday-school class. Pat followed me to Knoxville for his Ph.D.; today he's an anthropology professor at Cal State in Chico, California. *(Collection of Dr. Bill Bass)*

(RIGHT) A typical Arikara grave is circular. The round hole and flexed skeleton allowed a smaller grave—a big help to a gravedigger using a hoe made from a bison scapula.*(Copyright Doug Ubelaker; collection of Dr. Bill Bass)*

Steve Symes (left), Pat Willey, and I prepare to search the burned wreckage of James Grizzle's house outside Kingsport, Tennessee, in January 1981. *(Copyright David Hunt; collection of Dr. Bill Bass)*

We discovered the lower spine and left femur of James Grizzle beneath bricks and ashes in the basement of his house. The charred upper portion of his body and skull were found across the room, thrown by a dynamite blast. *(Copyright Steve Symes; collection of Dr. Bill Bass)*

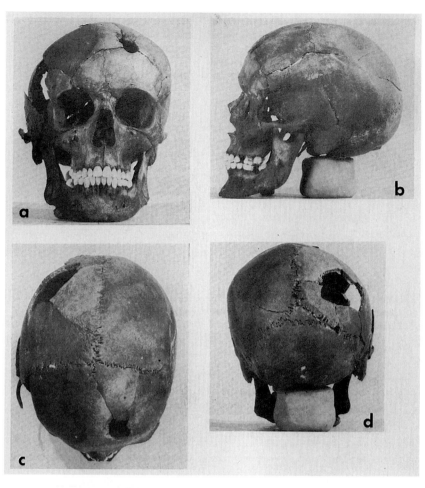

This shattered skull was found in Colonel William Shy's grave. A large-caliber bullet entered the skull above the left eye and exited at the back of the skull. *(Collection of Dr. Bill Bass)*

Colonel Shy was killed in the Battle of Nashville on December 16, 1864. These are probably the clothes he was buried in. *(Courtesy Carter House Museum, Franklin, Tennessee)*

As I removed human remains and clothing from atop Colonel Shy's coffin, I laid them out on a sheet of plywood on the ground. *(Courtesy Charlie Bass; collection of Dr. Bill Bass)*

This small family cemetery behind the antebellum Shy mansion near Franklin, Tennessee, was the setting for the forensic mystery that prompted me to create the Anthropology Research Facility—the Body Farm. *(Collection of Dr. Bill Bass)*

Bill Rodriguez, right, jokes with me during a break from catching flies in the course of his pioneering study of insect activity in four human corpses. *(Copyright Steve Symes; collection of Dr. Bill Bass)*

We built a small equipment shed at the Anthropology Research Facility, or ARF; progress must have been slow, judging by the inconvenient place I left my bag of nails. One Tennessee prosecutor joked that our decomposition lab should be called the *Bass* Anthropology Research Facility: BARF. *(Courtesy Jim Bass; collection of Dr. Bill Bass)*

Doorway to Death's Acre: The Body Farm's main gate consists of a razor-wire–topped chain-link fence, plus a high wooden privacy fence. *(Copyright Jon Jefferson)*

Our first research bodies were locked inside this chain-link enclosure; later, when we fenced the entire site, we expanded the research area to include studies of bodies on the ground, in shallow graves, and in other realistic settings. *(Copyright Jon Jefferson)*

This photo appeared in *The Knoxville Journal* in May of 1985, one day after the first (and only) protest at the Anthropology Research Facility. The picketers weren't protesting the Body Farm's decomposition studies, just its location. *(Reprinted courtesy Maryville-Alcoa Newspapers, LLC)*

I met Steve Symes in South Dakota in the 1970's and persuaded him to come to the University of Tennessee. Steve helped me with many forensic cases, wrote a pioneering dissertation on saw marks in bones, and went on to become one of the world's top experts on dismemberment. *(Copyright Hugh Berryman; collection of Dr. Bill Bass)*

One of our early experiments studied whether body bags are leakproof. They aren't. *(Collection of Dr. Bill Bass)*

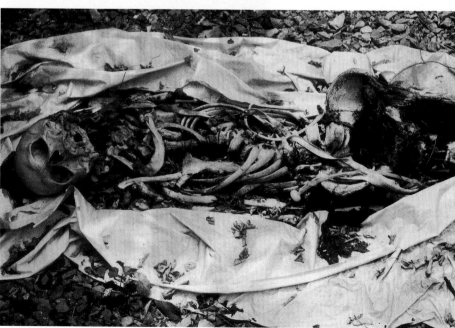

Patricia Cornwell visited Knoxville in researching her fifth novel, *The Body Farm*, in the early 1990's. The book's success gave our facility unparalleled exposure. Shown here in October 2002, she returned to consult me about insect activity in corpses. *(Copyright Jon Jefferson)*

Teeth and jaws can reveal much about an unidentified body: age, race, socio-economic status, even identity. This mandible came from an adult Caucasoid male whose body was donated to us for research study. *(Copyright Jon Jefferson)*

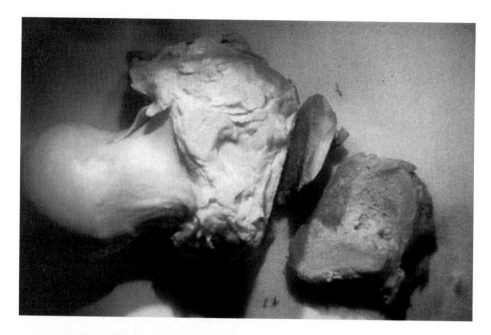

After examining this femur, taken from a teenaged murder victim, Steve Symes determined that a circular saw was used to dismember her body. The thin wedge of bone at the center resulted from a deep false start, right center, followed by a complete cut slightly to the left. Deep false starts are common in dismemberments with power saws. *(Copyright Steve Symes)*

Dismemberment expert Steve Symes relaxes at a forensic conference in Hawaii. *(Collection of Dr. Bill Bass)*

These bones came from the body that decayed undetected in a vacant lot adjoining Broadway Avenue in Knoxville. *(Collection of Dr. Bill Bass)*

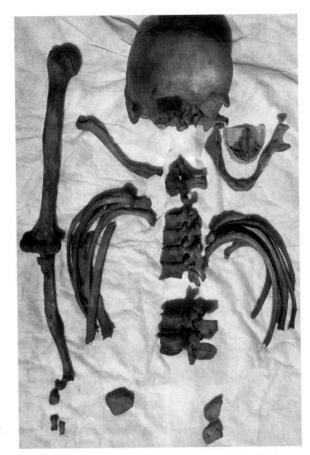

Two research projects at the Body Farm compared the rates of decomposition in the passenger compartment and the trunk of a car. Because car interiors can be hot and dry—and difficult for insects to access—bodies inside often mummify. *(Copyright Ed Richardson)*

This aerial view of the Body Farm, taken in October 2002 from Patricia Cornwell's helicopter, shows the gate, the clearing, and a decomposition experiment. The angular wooden structures are tripods used in weighing bodies; one corpse in this study of weight loss during decomposition shed forty pounds in one day. *(Copyright Jon Jefferson)*

We studied the decay of embalmed bodies in 1987. Judging by the disintegrating lower limbs, the undertaker skimped shamefully on embalming fluid in the lower body. *(Collection of Dr. Bill Bass)*

(LEFT) This charred Chevrolet Suburban in Monterrey, Mexico, contained incinerated human remains. The $7 million question was, were they the remains of Madison Rutherford? The vehicle's roof sags at its right-rear corner, indicating that gasoline or another accelerant was probably poured there. (*Collection of Dr. Bill Bass*)

(ABOVE) This cabin in Summit, Mississippi, was the scene of a family tragedy. Inside, Michael Rubenstein murdered Darryl, Annie, and Krystal Perry for the $250,000 life-insurance policy he had taken out on Krystal. Note the cabin's stout construction of stacked two-by-fours. (*Courtesy Mississippi Highway Patrol*)

Michael Rubenstein was convicted of killing his wife's son, daughter-in-law, and granddaughter and then leaving their bodies to decay in the cabin for a month. (*Courtesy Mississippi Highway Patrol*)

Art Bohanan holds a plastic bag in which I'll place a murder victim's severed right hand. Matching this hand's prints with the prints on an apartment lease, Art later managed to identify the victim as Darlene Smith. *(Copyright Bill Grant)*

Darlene Smith's body arrives at the Body Farm for cleaning and examination by, from left, Emily Craig, Lee Meadows, and Bill Grant. The body bag contained thousands of maggots but did not contain the hyoid bone, crucial evidence as to whether she'd been strangled or not. I sent a team back to find it. Their equipment rests on the hood of a car we used in our automotive decomposition studies. *(Collection of Dr. Bill Bass)*

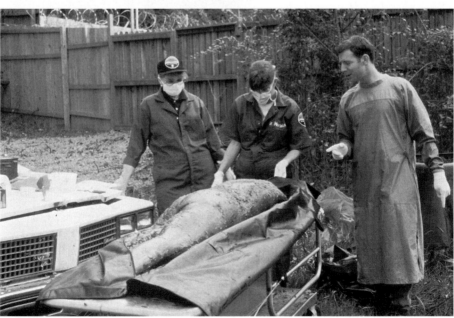

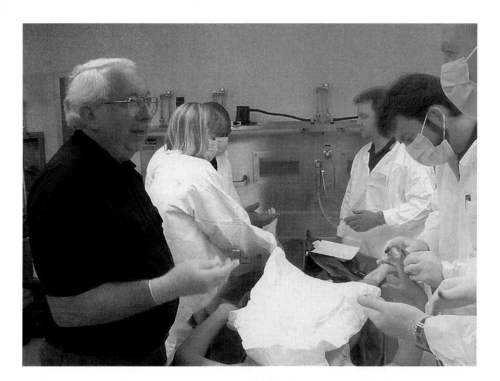

Art Bohanan is one of the nation's top fingerprint experts. Here he shows trainees at the National Forensic Academy in Knoxville how to fingerprint a corpse. *(Photo by Jarrett Hallcox; copyright National Forensic Academy)*

The body of Lloyd "Chigger" Harden was sent to Tri-State Crematory in Noble, Georgia, for cremation. Tri-State sent the family back a set of cremains, but two years later Chigger's body was found on the grounds of the crematorium. These are the burned, or calcined, bones resulting from the subsequent cremation of Harden's decomposed body. *(Copyright Jon Jefferson)*

Annette Blackbourne Bass, my second wife, with me on an Alaska cruise, around 1995. *(Collection of Dr. Bill Bass)*

(RIGHT) In 1998, I married Carol Lee Hicks, a close friend ever since my childhood in Virginia. *(Photo by Allyson Lauder; collection of Dr. Bill Bass)*

(LEFT) Lieutenant Ann Owen, my wife and partner for forty years, was an Army dietician. This photo was taken during her internship at Walter Reed Army Hospital in Washington, D.C., in 1951. *(Collection of Dr. Bill Bass)*

inches long. Most of the bones of the upper arms had been broken with violent force, and the legs had been cut from the pelvis at the hip joint.

This skeletal jigsaw puzzle just went on and on. But I was making progress, I reminded myself: In fitting these new pieces into the old puzzle, I found that one of the unburned fragments—the proximal end of the radius (the "elbow" end of one of the forearm bones)—fit perfectly with a burned piece of radius I'd received in Corporal Kelleher's first FedEx shipment. One of the new femur fragments fit perfectly with the femoral shaft recovered in the second batch, found in the woods behind the house. (That femoral shaft also yielded DNA that further corroborated the identification made from dental records.) So although some details remained confusing—*very* confusing—one thing had become crystal clear: All three sets of bone fragments, recovered from three different locations over the course of fifteen months, came from Sheilah Anderson, a woman whose husband claimed she had lit out for parts unknown.

Parts unknown, indeed. That wasn't *where* she ended up; that was *what* she ended up—or would have, if Kelleher hadn't been such a dogged investigator. The case was one of the strangest I'd ever encountered, and one of the most bizarre aspects was this: to all appearances, Jim Anderson was quite willing to murder and dismember his wife . . . but, by golly, he wasn't about to violate a municipal ordinance against unpermitted open burning! So he'd obtained the requisite permit authorizing him to burn trash on June 12, and we know for a fact that he had a fire in the yard on the appointed day, because Alexandria's fire chief drove past to make sure the blaze was under control.

Just imagine the scene: a murderous husband, burning his wife's body in the front yard, smiling and waving to the fire chief as he drives past. If a screenwriter pitched that story to a Hollywood film studio, he'd probably be laughed right off the lot. To Corporal Kelleher, though, and the prosecutor, Assistant Attorney General Janice Rundles, this was no laughing matter. Would a New Hampshire jury believe the bizarre scenario?

On the flight back to Knoxville, I racked my brain for any additional evidence that might be gleaned from the mutilated bones. I had already told Kelleher and Rundles everything I could. If anyone could tease additional clues from the severed, charred fragments, it would be Steve Symes, a former student and now a highly respected colleague. After I was back in Knoxville, I called Steve to propose a most unorthodox threesome: Could he slip away for a weekend in a secluded cabin with me and Sheilah Anderson? He could, he said. We agreed to rendezvous at Montgomery Bell State Park.

Montgomery Bell lies halfway between my office in Knoxville and Steve's morgue in Memphis, four hundred miles away. Rolling hills, covered with oaks and hickories, wrap around a lovely little lake, apparently brimming with fish (15-Inch Size Limit on Bass, cautions a sign at the water's edge). A six-story lodge of stone stood on a peninsula; half a dozen cabins perched halfway up a hillside, and ours was spectacular. The windows shed plenty of light on the dining table, where we laid out Sheilah Anderson's burned and fragmented bones. Murder investigation with a view.

Sheilah's dismemberment was one of the most complex and puzzling that Steve and I had ever seen. Judging by the fractures in the bones of the arms and legs, her limbs appeared to have been removed by blunt-force trauma. Her pelvis, ribs, and spine, however, seemed to have been cut apart with a wickedly sharp implement of some type.

Steve was immediately struck, as I had been, by the differential burning. The bones recovered from the front yard in 1993 were burned far more severely than either those recovered from the backyard shortly afterward or the skull and other bones found by the road crew in 1994. Steve suggested that the burning had occurred in two stages: Her entire body was put on the fire that the fire chief saw in June of 1993, he hypothesized. When that fire didn't do the job, the skull and other parts were removed and discarded—some in the backyard, some by the roadside—and the remaining bones were burned again in the front yard, this time more thoroughly.

Burning had destroyed the tool marks on the first set of bones; however, the unburned and slightly burned portions of the skeleton gave Steve some undamaged marks to study. Unlike many dismemberment cases, these bones showed virtually no traces of false starts, hesitation marks, or interrupted cuts. The tool marks indicated that the bones had been sliced or shaved decisively, sharply, and forcefully. The cuts were made not by a sawing motion but by a chopping motion, and they were made with enough force to cut through some of the bones with a single blow. The blade was sharp enough to shave off thin sections of bone in spots—for example, a slice from the body of one of the vertebrae—yet heavy enough to cut through such large structures as the hipbones and the femora.

Steve and I were both puzzled by the case. The marks on the cut faces of the bones were odd too. They indicated a blade that was curved; however, that in itself wasn't the odd part: many common garden tools have curved blades. Whatever this implement was, though, it had a tighter curve to the edge than any ax or shovel we'd ever seen. If the curve or arc of the edge were extended to form a complete circle, that circle would have measured less than three inches in diameter. Considering the great force required to shear the bones, we wondered if the tool might have been a posthole digger chopping downward with a man's full weight behind it—but the edges of a postholer weren't that curved, either.

We spent all Saturday morning and half the afternoon studying and restudying the cut marks, considering and rejecting different tools as the implement of dismemberment. Then, in late afternoon, there was a knock at the cabin door. When I opened it, I found myself face-to-face with a park ranger. *Uh-oh*, I thought, *this means trouble*. Using my body as a screen, I tried to block the ranger's view of the bones spread all over the dining-room table.

The ranger's visit did mean trouble, though not because we were using the cabin as a forensic laboratory. I'd gotten a phone call from Knoxville, said the ranger, and it sounded urgent. Leaving Steve with the bones, I hurried up to the lodge. The call was from Dot Weaver, a friend who was caring for my ninety-five-year-old mother; when I returned the call, she told me that Mom had just suffered a series of small strokes and had been taken to the hospital.

I told Steve we needed to cut our work short. There wasn't that much more he could tell me anyway, he said. We took one last look at Sheilah Anderson's ravaged remains, hoping that Janice Rundles, the New Hampshire prosecutor, wouldn't have to depend solely on our meager findings to cinch the case against Jim Anderson. Luckily, she didn't: just before the case came to trial, Anderson—once one of New York City's Finest—pleaded guilty to murdering his wife. Shortly after he was imprisoned, he took a guard hostage, held him for several hours, and beat him severely. Maybe someday he'll say what it was he used to chop up his wife's body.

My weekend rendezvous with Steve hadn't been quite as satisfying as I'd hoped it would be, but that's the way some cases go: All you can do is look at the evidence and listen to the bones. The bones don't always tell you the whole story, but when they do, the tale can be both horrifying and hypnotizing.

Steve found that out firsthand from a victim named Leslie Mahaffey. . . .

I FIRST MET STEVE a quarter-century ago, out in the boonies of western South Dakota. He was a scrawny twenty-four-year-old with a B.A. in anthropology; after graduation he'd gotten a job cataloging bones for Bob Alex, South Dakota's state archaeologist. Steve's main task was to sort and catalog thousands of Sioux and Arikara Indian bones from the W. H. Over Collection, assembled by a self-taught South Dakota archaeologist during the late 1800s and early 1900s.

In 1978, in one of the first large-scale repatriations of Native American remains, Bob Alex persuaded the government of South Dakota to return the bones in the Over Collection to the Arikara and Sioux tribes for reburial. Before giving the bones back, though, he offered to let me study them a while.

The collection was housed in a former military hospital northwest of Rapid City. Late in the spring of 1978, I arrived from Knoxville in a Ford station wagon, towing a U-Haul trailer that would carry the collection back to Tennessee. Steve had been under the gun to complete his inventory and box up the bones before I got there. On his desk I saw an open and well-thumbed copy of my guidebook to bones, *Human Osteology: A Laboratory and Field Manual*. (Since it came out in 1971, the book has gone through twenty-three printings and sold somewhere around seventy-five thousand copies, making it something of a blockbuster among textbooks, I'm proud to say.)

We howdied and shook hands. "I see you're using my book," I said.

"Well, I tried some others," he said, "but this is the only one that's actually helpful in identifying some of the more difficult bones."

Clearly this was an exceptionally bright young man. Quite possibly a genius.

Within ten minutes of meeting Steve, I realized—and not just because I was flattered by his comment—that he had the makings of an exceptional anthropologist. He was knowledgeable and curious, but clearly mature, disciplined, and steady, too—a combination far less common in would-be professors than you might think. Unlike a lot of today's students, he hadn't swallowed some romanticized image of anthropology from TV shows or Hollywood movies. He knew it took a lot of hard work, and he seemed more than willing to get dirt under his fingernails. By the time we finished loading up the U-Haul, I'd convinced Steve that he should go to graduate school, and I'd made a pretty strong case for Tennessee as the place to do it. There was one minor problem with that plan, though. Our graduate slots for the coming fall were already filled.

Four months later, Steve showed up in Knoxville anyhow, the academic equivalent of a football walk-on, hoping to snag a place on the starting team a week before the opening game. I shoehorned him into classes in archaeology and osteology, hoping that a slot in the forensics program would open up soon—and that Steve would still be interested by then.

It did, and he was. He quickly absorbed the rest of my osteology handbook, thereby earning a place on our forensic response teams. The walk-on had done it: He'd made the anthropological equivalent of varsity first string. In the field, Steve showed a quick grasp of crime scene investigation. Equally important, he was a superb photographer. When it comes to crime scene photos, more is always better, and great is always best. Steve's crime scene photographs were—still are—the best I've ever seen.

After eight long years of graduate work and crime scene assistance, Steve passed his doctoral examinations, then took a job as the staff forensic anthropologist for the medical examiner in Nashville. Besides working full-time for the ME, Steve planned to research and write his Ph.D. dissertation there in Nashville. His topic: estimating age by examining the sternal end of the clavicle. ("The collarbone's connected to the . . . breastbone. . . .")

Then came another one of those pivotal points in Steve Symes's life. During the course of a few violent weeks in Nashville, Steve was handed three dismemberment cases. The detective working one of the cases pointed to a notch in a bone and asked Steve to tell him about it. Happy to have a chance to display his expertise, Steve drew himself up and said, in his most professorial voice, "Why, that's a saw mark in an arm bone."

The cop stared at Steve in disgust. "I *know* it's a saw mark in an arm bone," he snorted. "You're the bone doc; what *kind* of saw mark?"

Steve didn't know, but after he finished blushing, he decided to find out—not just about that particular saw, but about all types of saws.

At this point I can't help but add that I'd been trying for years, unsuccessfully, to interest a graduate student in researching saw marks. We'd had a sensational dismemberment case in Knoxville in the mid-

1980s. A love triangle turned hateful, and the woman and one of her men ended up killing her other man, then cutting him up and scattering the pieces all over town. That case had gotten me to thinking about how little we knew about what evidence might be left behind by a saw as it cut up a body. But nobody seemed inclined to pursue the topic, including Steve, until that bloody Nashville summer when he found himself butting up against the problem not once but three times.

Police departments and courts all over the world have long considered ballistics evidence to be scientifically credible. Just like people, guns leave fingerprints: A pistol's firing pin makes a consistent impression in every cartridge it strikes; the rifling in the barrel leaves characteristic grooves on each slug that spirals toward a victim; the ejector mechanism scratches or dents every spent shell case the same way as it kicks it out of the breech.

If guns leave telltale marks, why wouldn't saws? Steve and I felt sure they would. At the time, though, we seemed to be in the minority. Conventional wisdom held that every stroke, every pass of a saw erased the marks left by the previous stroke; in other words, saws covered their own tracks. Steve made up his mind to prove that they didn't—that there was a world more detail to be seen, a world more evidence to be gathered.

Over the next two years Steve bought or borrowed every kind of saw he could lay his hands on: ripsaws, crosscut saws, hacksaws, jigsaws, coping saws, circular saws, chop saws, Japanese pull saws, and more. He spent several weekends with Dr. Cleland Blake, an East Tennessee medical examiner who was also a master woodworker—and studied hundreds of saw blades in Cleland's collection, ranging from jewelers' trim saws to lumberjack-grade chain saws.

Clamping donated arm bones and leg bones in his bench vise, Steve made thousands of experimental cuts and studied them through microscopes. At first, he saw little that seemed meaningful. Eventually, though, he found the key. Peering through a surgeon's operating microscope and angling light across the cut marks, he saw a world of three-dimensional detail open before his eyes: immense canyons and jagged cliff faces carved in bone. He took countless micrographs, plaster impressions, and measurements, cataloging push strokes, pull strokes, rotary cuts, false starts, skips, hesitations, and other telltale marks left by saws as they ripped through the bones.

I'll never forget the first time Steve hauled me into a lab, led me to a stereo microscope, and gave me a stroke-by-stroke replay of the saw marks in a femur he had clamped in a vise and sawed in half. Etched forever in a cross section of bone—as they are now in my mind's eye as well—were the zigzag bite marks left by individual teeth sliding back and forth, chewing their way relentlessly downward through the bone in a series of shallow, **Z**-shaped tracks. It was a moment that made me proud and humble at the same time: The student—*my* student—had surpassed the teacher, in at least this one macabre specialty.

Eventually, Steve was able to look at a bone fragment from a murder and see far more than "a saw mark in an arm bone"; eventually he was able to discern, for instance, the tracks of a ten-teeth-per-inch crosscut saw, with a kerf (cut) width of 0.08 inch, created by alternating offset teeth, cutting on a push stroke—a stroke interrupted, he might observe, by three skips, two false starts, and one temporary halt. A husband cutting up his wife's body wouldn't mean to leave such telltale tracks behind, any more than a hired shooter means to leave ballistic evidence on his bullets. It's simply the unavoidable consequence.

Steve never got around to writing that serviceable, boring dissertation on the sternal end of the clavicle; instead he wrote *Morphology of Saw Marks in Human Bone: Identification of Class Characteristics*—despite the dry-sounding title, a unique and pioneering contribution to forensic anthropology and homicide investigation.

Not long after he began his research on saw marks, Steve moved westward again, this time to Memphis. Word of his gruesome specialty spread; gradually, packages of dismembered body parts began arriving in Memphis from other cities and states and countries, shipped to Steve by police or prosecutors desperate to narrow the search for a killer or a murder weapon. His most sensational case began on April 6, 1992, when Mike Kershaw, a Canadian police constable, called to ask for Steve's help with a gruesome killing that had occurred the previous June in Saint Catherines, a midsize city located across the tip of Lake Ontario from Toronto.

Leslie Mahaffey, a fourteen-year-old Saint Catherines girl, had stayed out late one night with friends, missing her 11:00 P.M. curfew by several hours. Sometime around 2:00 A.M., as she walked alone from a phone booth toward her house, she was abducted. Two weeks later, fishermen discovered her body. It had been cut into ten pieces, encased in blocks of concrete totaling 675 pounds, then dumped into two nearby rivers. The blocks were exposed when the water level dropped by several feet. Leslie's brutal murder had terrified the public and baffled the police; Constable Kershaw hoped Steve might be able to shed light—any light, no matter how faint—on the killing or the killer.

On April 30, Kershaw came to Memphis, bringing Leslie Mahaffey's butchered bones: sections from both femurs, both upper arms, two lower arm bones, and two cervical vertebrae. The specimens had been immersed in formalin to preserve them. Despite the passage of nearly a year, the bones still had soft tissue on them.

The very day Kershaw arrived in Memphis, another Saint Catherines girl, Kristen French, was found murdered; it appeared that she'd been raped and sexually tortured before she was killed. Canadian police knew that if they didn't catch the killer soon, more girls were likely to die horrible deaths.

Steve began by photographing each bone, then he cooked them in hot water for several hours and gently teased away the soft tissue. Right away he could tell that all the cuts were from the same type of saw. The cuts were very uniform; the cut surfaces were smooth, almost as if they had been polished, and there was very little breaking or chipping at the places where the saw had entered and exited each bone.

There were, however, numerous false starts, places where the saw began cutting into the bone, but—maybe because the position or angle was awkward, maybe because the bloody saw slipped from the killer's grasp—the blade had jumped out of the groove, forcing a new cut to be started. Several of the false starts were quite deep, extending almost all the way through the bone. That told Steve that the cutting was easy—the mark of a power saw of some type—because if you're using a hand-saw, and the saw jumps out of a deep groove, you don't start a new one, you maneuver the blade back into the groove you've already cut. The deep false starts—together with the uniform width of the grooves, the polished-looking surfaces, and the convex curve of the cut marks—told Steve that Leslie's body had been cut up with a circular saw with a blade diameter of 7¼ inches or greater.

Of course, a lot of Canadians own circular saws; Steve could tell the police what type of saw had cut up the body, but he couldn't tell them whose garage or basement to search for it. The case remained unsolved for ten more months. Finally, in the winter of 1993, the police got a huge break. A twenty-three-year-old woman named Karla Homolka came forward with a sordid and shocking story. Her husband, a bookkeeper named Paul Bernardo, had abducted Leslie Mahaffey and Kristen French to use as sex slaves, she claimed. Paul had forced Karla to participate in some sex acts, she said, and to videotape others. After an escalating series of degrading and violent acts, he'd strangled the girls. Besides Leslie Mahaffey and Kristen French, there was a third victim, Karla claimed: her own younger sister, Tammy, whom Paul had drugged and raped back in 1990. While still unconscious, Tammy had vomited and choked to death; until Karla's visit to the police, her sister's death had been considered merely a tragic accident.

On the morning of Monday, June 12, 1995, Steve walked up the steps of a Toronto court building to testify in Paul Bernardo's murder trial, which had begun four weeks earlier. Canadian reporters were fascinated by Steve and his grisly specialty. "You'll never meet another man like this," began one newspaper story, "and you probably won't mind." The story went on: "As far as he knows, he is the only person in the world to have earned his doctorate by using bones to differentiate the tools used to tear apart human bodies."

When Steve was called to the witness stand by the frock-coated crown prosecutor, he painted a precise and horrifying picture of Leslie Mahaffey's dismemberment. The kerf width—the groove cut by the saw blade—was unusually narrow in Leslie's bones, indicating a thin blade. Most carbide-tipped circular saw blades leave a kerf of about ⅛ (0.125) inch; the blade that butchered Leslie was thinner, with a kerf width of just 0.08 to 0.09 inch. When Steve made experimental cuts of his own in other bones, using circular saws with blades ranging from 7¼ to 12 inches in diameter, he testified that his cuts were more uniform, showing less tendency to drift, than the cuts in Leslie's bones. But Steve had an advantage that Leslie's killer had not had: he was cutting clean, dry, defleshed bone, and it was rigidly anchored in a vise.

On cross-examination, Paul Bernardo's attorney asked just one question: Would cutting up a body with a circular saw make a mess? A big mess, Steve answered. The courtroom crowd was horrified by what Steve had to say, but their horror was lessened by the way he said it—by what one reporter described as his "open American manner and a self-deprecating sadness." Self-deprecating is right: Steve is one of the world's top five experts on tool marks in human bone, but he's remarkably modest and unpretentious.

When Paul Bernardo took the witness stand, he denied that he'd murdered Leslie Mahaffey; he claimed both she and Kristen French had died accidentally while he was out of the room. He did, however, admit to dismembering Leslie. He cut up her body, he said, with an old McGraw-Edison saw: a circular saw of the type described by Steve. In fact, the saw, which Bernardo had gotten from his grandfather, was found in the basement of his tidy bungalow in a Saint Catherines suburb. Unfortunately for the prosecution, the blade and part of the housing were missing.

Steve left Toronto the day after he testified, hoping he'd done some good, but juries are funny: you can never tell for sure what's going to hit home with them. Bernardo's trial dragged on through June, through July, into August. Then, as the trial was nearing its end, a dramatic flourish made fresh headlines: The crown prosecutor concluded his case by producing a rusted saw blade that a police diver had fished from the lake only days before. Alongside the blade, the diver also found part of the housing of a power tool. The blade and the housing fit Bernardo's old McGraw-Edison saw perfectly. The blade also fit Steve's cut-mark analysis to a T: a circular saw blade, 7½ inches in diameter, thinner and finer-toothed than most modern, carbide-tipped blades, with the right width to have made the 0.08-inch cuts.

Paul Bernardo was convicted on two counts of murder and sentenced to two twenty-five-year prison terms, without the possibility of parole. I'm told he gets fan letters and phone calls from teenaged girls. I know a lot about human bones, and so does Steve Symes. But there's a lot more we'll never comprehend about the dark recesses of the human heart.

CHAPTER 14

Art Imitates Death

B Y 1993, I'd been running the anthropology department at the University of Tennessee for more than
two decades. I'd helped create a new Forensic Anthropology Section within the American Academy of Forensic Sciences (AAFS), an important milestone in the
development of this fascinating new field. I was also serving my twenty-second year as Tennessee State forensic anthropologist, a position that led to interesting forensic
cases in nearly all of Tennessee's ninety-five counties. My
relationships with police departments, district attorneys,
the TBI, the FBI, and other law enforcement agencies
were strong. I lectured frequently to groups of medical examiners, doctors and dentists, police, and funeral directors. I testified in court several times a year; occasionally I
made it into the newspapers or onto the television news,
especially if there was a particularly gruesome case or if I

won some teaching award, as I did in 1985, when the Council for the Advancement and Support of Education did me the honor of naming me National Professor of the Year. All in all, I thought, things were about as busy and exciting as they could possibly be.

I couldn't have been more wrong. One brief phone call, and things would intensify beyond my wildest imaginings. For quite a few years I'd been lecturing regularly at forensic meetings all over the country. At one of those meetings I met a young assistant medical examiner from Virginia, Dr. Marcella Fierro. Over the years, seeing one another at meeting after meeting, we became good friends. Eventually, after Dr. Fierro became Virginia's chief medical examiner, she began inviting me up to lecture to her staff once a year, either to broaden their horizons or just to strengthen their stomachs.

Most medical examiners are forensic pathologists—physicians specializing in disease or trauma to tissue. If they're able to autopsy a body within a few hours or even a few days of death, they're often remarkably successful at determining time since death and cause of death. But once decomposition reaches a fairly advanced stage, an autopsy becomes difficult. The soft tissues begin to liquefy through a combination of bacterial action, cellular chemical changes (a pH disruption called autolysis), and maggot feeding. As the soft tissues disappear, so do the physical clues a pathologist looks for, such as knife wounds in flesh. But if there are knife marks or other types of bone trauma, a skilled forensic anthropologist can often deduce an amazing amount of information from the skeleton long after an autopsy is impossible.

In 1984 a young technical writer had joined Dr. Fierro's staff in Richmond. The woman, a former crime reporter, was clearly very intelligent, highly articulate, and fascinated by forensic investigations. She was also an aspiring crime novelist. After six years in Dr. Fierro's office, she sold her first mystery novel.

That young woman's name was Patricia Cornwell, and that novel, *Postmortem*, established her as a remarkably talented crime writer. It won five major international awards the year after it was published, and it remains the only mystery novel ever to do so. *Postmortem* marked not just Patricia Cornwell's debut but the debut of her recurring heroine, Virginia medical examiner Kay Scarpetta. Dr. Scarpetta was tough on the outside, tender and wounded on the inside. She could have been inspired by Cornwell's boss and mentor, Dr. Marcella Fierro, in her professional life, and Cornwell herself, I suspected, in her personal characteristics. In any case, Scarpetta swiftly became one of the most charismatic superstars of crime fiction. So did Patricia Cornwell.

Patricia Cornwell and I first met at one of Dr. Fierro's annual training seminars, while she was still with the medical examiner's office. As usual, I was showing slides of maggot-covered bodies. She introduced herself afterward, asked a lot of questions about my research, and complimented me on my presentation. End of story—or so I thought.

Then, in the summer of 1993, I got a phone call. The voice at the other end of the line said, "Dr. Bass, this is Patricia Cornwell." She reminded me who she was and where we'd met—by now she was rich and famous, and no longer working for Dr. Fierro—then she came straight to the point: "I was wondering if you might be willing to run a little experiment for me at your research facility." She was working on a new novel, she explained, in which she planned to have the killer return to the death scene—the basement of a house—some days after the murder and move the body to another location. What signs or marks, she needed to know, could a body pick up as it began to decay, and how much of that detail would remain once the body was moved to a new location?

This was a first. I'd been asked to study particular phenomena by medical examiners and homicide detectives, but never by a novelist. My first inclination was to say no, but as she described what she had in mind, my scientific curiosity was piqued. These were interesting questions. By now we'd been studying decomposition at the Anthropology Research Facility for a dozen years, but up to this point, most of the bodies had been buried or simply lying outdoors on the ground. Our main research focus had always been to learn more about the processes and timetable of decomposition so that we could help law enforcement estimate time since death more precisely and accurately. Cornwell's request opened up a whole new research area.

I called Detective Arthur Bohanan, my friend and colleague at the Knoxville Police Department, to get a homicide detective's perspective on whether this sort of experiment sounded helpful and what kinds of information would be most valuable. Art was not your average cop. Over the years he had turned himself into a true expert on fingerprints—specifically, on ways to capture them from surfaces that had never yielded prints before: fabrics, paper, even the skin of a murder victim. He'd gone so far as to patent an apparatus that would vaporize cyanoacrylate—superglue—and waft it across surfaces or throughout an entire room. If you've ever accidentally superglued your fingers together, you know how eagerly the stuff bonds to human fingertips. Art figured out that it also bonds to the oils that fingertips leave on things they touch. His apparatus, which is now used by crime technicians worldwide, can capture latent fingerprints that routine dusting could never reveal. Recently the FBI ordered another sixty-six of Art's machines; for a fingerprinting system, you can't get a better product endorsement than that.

As we talked about the experiment Cornwell wanted me to run, Art became more and more enthusiastic. If a fingerprint on a body could help crack a case, why not some other distinctive mark? He'd seen odd imprints and discolorations on bodies before but didn't have any data that could help explain them. That settled it: I would do the experiment. Together, Art and I called her to discuss the setup in more detail.

Cornwell planned to set the murder in a basement in the town of Black Mountain, North Carolina. One of the trademarks of Cornwell's fiction is her frequent use of places she's been or experiences she's had. Black Mountain is a summer resort town where she spent much of her youth. North Carolina and Tennessee occupy roughly the same latitude and share a border, defined by the crest of the Great Smoky Mountains. Black Mountain lies about the same distance east of the crest that

Knoxville lies to the west of it, so the climate at her crime scene closely resembled the climate at our research facility.

To simulate a basement, we'd need a concrete slab. Coincidentally, that part of the experimental setup was already prepared: We were just about to build a storage shed at the research facility for gardening tools, medical instruments (the scalpels and other implements needed to cut apart a skeleton at the end of a research study), and a small weather station; as a first step we'd recently poured a slab that would be plenty big for the experiment. To simulate an enclosed basement, all we had to do was build a "room" atop the slab—basically, a simple plywood box measuring eight feet long, four feet wide, and four feet high.

Then Bohanan and I realized we might have a problem. Summer was approaching fast, and the summers in East Tennessee are hot and muggy, with temperatures frequently ranging into the low to mid-90s— quite a bit warmer than a below-grade basement in Black Mountain would be. We called Cornwell to discuss the problem; she told us to buy an air conditioner, if that would resolve it, and to send her the bill. We needn't have worried. There are ebbs and flows in the donated-cadaver business, and that summer, for some reason, we hit a slow period. Before long, summer was over; football season and fall weather arrived.

So did Patricia Cornwell. In September of 1993, on a football weekend, she paid us a visit. Football weekends in Knoxville are crazy; she booked what was probably the last available hotel room in the city, and she dined amid throngs of orange-clad UT fans at a popular riverfront restaurant near the stadium. I took her out to the research facility, where she took copious notes as I showed her corpses in various stages of decomposition and explained some of the graduate students' research projects.

A few weeks later Arthur Bohanan and I took a donated corpse's fingerprints, then we drove him—corpse 4-93—out to the facility. Together we wrestled the body out of the truck and into our plywood box. We positioned the body on its back, as Cornwell had requested. Beneath it we placed a coin—a penny, lying heads-up—along with a key, a brass strike plate from a door frame, a pair of scissors, and a chain-saw chain. Then we closed the door and walked away, just as the killer in Cornwell's story was going to do.

Six days later we went back, dismantled the box, and retrieved the body. However, unlike Cornwell's killer, who dumped the victim's body beside a lake, we took ours to the morgue so we could examine and document whatever traces or clues the simulated death scene might have left. Imprinted on the body's lower back was a perfect circle. Within the circle a faint imprint of Abraham Lincoln's head was clearly visible. The imprint was not quite as distinct as what you'd get if you put a piece of paper on top of a penny and rubbed a pencil lead across it, but it was amazingly close. The disk was brown with specks of green—copper oxide from the penny's corrosion by body fluids.

The key and the strike plate were sharply outlined on the legs. So was the pair of scissors that we'd placed under the back; its handles left perfect ovals in the flesh. The chain-saw chain left a sinister, coiled imprint, discolored a deep reddish brown along the teeth, almost as if they had bitten into the skin.

The body bore one other mark as well: a distinct, raised line of flesh zigzagging across the back and shoulder. That one was a puzzle to us at first; then we took a closer look at the spot where the body had lain. Running through our concrete slab, which had been poured by rank amateurs—namely, me and my students—was a crack whose zigs and zags matched those on the body perfectly.

Arthur and I were both delighted with the results; so was Patricia Cornwell when we sent her a research report and copies of our photographs. She said the experiment had given her exactly the kind of detail she needed for her book.

The next time I saw Cornwell again was the following February, at the annual meeting of the American Academy of Forensic Sciences in San Antonio, Texas. As a crime writer, she was always on the lookout for new techniques that might make her books more interesting and realistic, and the AAFS meetings were often the place where researchers unveiled scientific breakthroughs and new forensic technologies. I bumped into her on a balcony overlooking the lobby of the Marriott River Center, the hotel where the conference was taking place. I asked how the book was coming along; she said that it was finished and that she was quite pleased with it. She thanked me again for running the experiment, then she added, "I'm calling the book *The Body Farm*." You could have knocked me over with a feather.

When we first began researching human decomposition back in 1980, our facility didn't even have a name. After all, it was really just a two-acre patch of ground, fenced off to keep out carnivorous animals and curious humans. The original fence was chain-link, but after a few passersby caught traumatic glimpses of the bodies inside, we added a wooden privacy fence. At some point, probably when we began writing up our research results for scientific journals like the *Journal of Forensic Sciences*, we decided we should probably call it something scientific-sounding. So we named it the Anthropology Research Facility, or ARF. Well, it wasn't long before some wag with the local district attorney's office suggested renaming it the *Bass* Anthropology Research Facility, or BARF. Luckily, that nickname never caught on; instead, police and FBI agents gradually started referring to it as "the Body Farm." Before long, I was calling it that too. It's easier to say and a lot more descriptive than "Anthropology Research Facility."

When Cornwell asked us to stage the experiment for her, I had no idea the facility itself would figure in her book; I assumed she'd use some of the research data, and that would be it. Instead, here she was telling me we were the title attraction. I was terribly flattered, and here's why: In all the years we'd been studying decomposition, nobody much had seemed to give a damn about our research—a few anthropologists and entomologists, maybe, but that's about it. Then along comes a famous writer who wants to name her book after our facility. What a nice pat on the back! I told her I couldn't wait to read it.

A few months later a copy arrived in the mail. As I read it I was stunned. The research facility was featured, and glowingly; so was its director, "Dr. Lyall Shade." It was as if the world's biggest spotlight had just swiveled in our direction: The phone didn't stop ringing for weeks. Our departmental secretaries fielded dozens of calls from reporters asking for the Body Farm's number. There wasn't a phone out there in the woods, of course, but after the first hundred or so calls, I jokingly told the secretaries to tell the callers to hang up and call "1-800-I AM DEAD."

By 1996, *The Body Farm* was one of the best-selling mysteries ever published. The book was an international hit, selling hundreds of thousands of copies in England, Japan, and other countries. Someone I know was traveling to Japan regularly on business at the time, and he told me that his colleagues in Japan made him stuff his suitcase with copies of the book every time he came from America.

It wasn't long before a parade of reporters and television crews was beating a path to Knoxville and the Body Farm. Even now, some ten years later, the parade still hasn't stopped. Some of the stories have been lurid or laughable, but others have been factual and respectful.

But flattering as the attention was, it was also distracting. If we'd been willing to give up research, teaching, and writing, we could have devoted twenty-four hours a day to giving tours of the facility. I give around a hundred lectures a year to police, undertakers, ATF agents, and other groups, and nearly everybody I talk to asks to come to the Body Farm. One week, den mothers from two different Cub Scout dens called, asking me to take their kids on a tour of the Body Farm. At that point I finally snapped: clearly, things had gotten out of hand. I began saying no far more often than I said yes. And yet, I still say yes, and my colleagues still say yes, many times.

And some of the attention is a blessing. Because of Patricia Cornwell's blockbuster novel and all the subsequent media attention it sparked, we get far more calls than we used to from people who want to donate their bodies for research. What nearly all of these donors say when they contact the university is "I want to donate my body to the Body Farm."

In November of 2002, Patricia Cornwell published a remarkable new book—nonfiction, this time. Titled *Portrait of a Killer: Jack the Ripper, Case Closed*, it represented the culmination of two years of painstaking forensic research. In a case of life imitating art—or, more precisely, art inspiring life—the crime novelist has reinvented herself as a real-life forensic detective. Digging deep into the past and using up-to-the-minute DNA technology, her book makes the case that Jack the Ripper was a Victorian artist named Walter Sickert, who painted a gruesome series of murder pictures that bore striking resemblances to the murder scenes where Jack the Ripper left his victims. If Patricia Cornwell ever decides to give up fiction for good, the real world could use a tenacious forensic investigator like her.

There are moments in life when, in hindsight, you realize everything has changed forever. I'm proud to say the publication of *The Body Farm* was one of those moments in my life, and in the life of the Anthropology Research Facility I created. And I'm proud to call Patricia Cornwell both my colleague and my friend.

CHAPTER 15

More Progress, More Protest

SIX MONTHS after Patricia Cornwell's novel *The Body Farm* thrust the Anthropology Research Facility into the limelight, I was still basking in the glow of media attention. I'd always gotten along well with journalists, mainly because I didn't mind telling them what I learned when I examined decomposing bodies or bare bones. My openness had caused me some embarrassing moments—especially when I misjudged Colonel Shy's death by almost 113 years—but it had also helped educate the public about forensic anthropology and the role it could play in fighting crime.

By this time I'd been heading the anthropology department at the University of Tennessee for nearly twenty-five years. During that quarter-century, the faculty had grown from six to twenty. Our program had grown from a small undergraduate major to one of the nation's leading training grounds for forensic anthropologists: There were around sixty board-certified forensic anthropologists in the United States by now, and I'd helped train a third of them.

The Council for the Advancement and Support of Education had named me Professor of the Year, not just for UT or Tennessee, but for all of the United States and Canada. Not long after that, President Ronald Reagan came to Knoxville and had lunch with me. Our work was attracting recognition and acclaim, in America and around the world. I was invited to lecture in Australia, Canada, and Taiwan.

Much to my surprise, my personal life was full and happy again, too. The reason for that change had been right under my nose for twenty years. Ever since I moved to Knoxville to run the anthropology department at UT, I had loved going to work each day. One reason was the work itself: teaching is fun, mostly, and forensic cases are fascinating. Another reason was Annette Blackbourne.

I had hired Annette not long after I came to UT. The department already had one secretary, but as we expanded and began building a research program, we needed someone to keep track of our research grants. When I interviewed Annette for the job, I was impressed by her organizational and financial skills; I was even more impressed by her warmth, maturity, and empathy with people. In a large department like ours, populated by everyone from homesick first-year undergraduates to tenured, self-important professors, diplomacy and humor were crucial.

When our main departmental secretary left for a higher-paying job, I promoted Annette into that position; later still, her job was upgraded from secretary to administrative assistant. Perhaps *counselor* or *adviser* would have been a more accurate title. Whenever I faced a difficult decision, I ran it past Annette, and more than once she saved me from making a terrible mistake. For instance, when picketers showed up at the Body Farm, she kept me from rushing over to confront them. Instead we watched them, unnoticed from a car across the parking lot, chuckling at the cleverness of their protest banner; as a result I was able to respond to news reporters later with a much cooler, clearer head.

In twenty years of working together, Annette and I had never spoken a cross word to one another. Everyone in the department—the other faculty, the graduate students, the undergrads—adored her. Over the years Ann and I had become close friends with Annette and her husband, Joe, a pharmacist at UT Medical Center. Twice a year the four of us would pile into a car or a camper for a long weekend excursion somewhere in the Southeast: Nashville, Asheville, Chattanooga, Mammoth Cave, and half a dozen other destinations. Then, shortly before Ann got sick, Annette's husband was diagnosed with lung cancer. He died about the time Ann's cancer was diagnosed.

Throughout Ann's illness Annette was a generous and sympathetic listener, and when Ann died, she understood exactly what I was going through. Annette's friendship and understanding pulled me through those difficult first months; eventually that friendship deepened into love. Fourteen months after Ann's death, Annette and I got married in a small chapel at Second Presbyterian Church. I felt reborn. I felt young all over again.

Everything, in short, was going well in the fall of 1994. Too well to last.

Once again, the trouble began with waterlogged bodies. Years before, there had been that Roane County floater I'd stashed in the department's mop closet, provoking the wrath of the janitor. This time the problem started with Tyler O'Brien's adipocere study. Adipocere is the greasy, waxy substance that often coats bodies pulled from lakes, rivers, and damp basements. With all the water in Tennessee, I was quite familiar with adipocere. But, as usual, I didn't just want to know the *what* and the *why* of it; I also wanted to know the *when* of it, so that the next time a sheriff's deputy or rescue squad brought me a floater, I could look at the degree of adipocere formation and tell them, with at least some measure of scientific confidence, how long that body had been "sleeping with the fishes."

I'd tried to persuade several graduate students to do a master's thesis study on adipocere, but I hadn't found any takers; I guess they'd all been around long enough to know that floaters are corpses at their worst—their smelliest and slimiest. But finally, in the fall of 1993, along came Tyler O'Brien, who had spent the previous summer working for the medical examiner in Syracuse, New York. Syracuse is surrounded by New York's Finger Lakes, so Tyler saw quite a few drowning victims during his summer with the ME. Some of those drowning victims were adipocere-covered and others weren't, and Tyler, like me, was curious about the difference in conditions and time since death.

The simplest procedure would have been to moor bodies in the river below the research facility. But we didn't want fishermen calling the police every day for six months, so Tyler came up with a new system: he dug three grave-size pits in the ground, lined them with heavy plastic, and filled them with water. Tyler's narrower, more controlled study had a strong scientific argument in its favor. By limiting the number of variables—in other words, by factoring the possibility of any hungry fish out of the equation—he could focus purely on adipocere formation, without outside interference.

Tyler's study involved three bodies, one in each pit. To make it easy to study a body at various intervals during the experiment, he put a wire platform in the bottom of each pit and attached hooks to each corner so we could hoist it up; then he put the body on top.

The first body floated like a cork. We'd push his head down, and his feet would bob to the surface; we'd push his feet down, and his head would pop back up. We discussed weighting him down, but decided to let his body seek its own level in the water. The second body sank like a rock. Often drowning victims or murder victims thrown in a lake or river will rise to the surface after a few days or weeks—when enough decomposition gases have built up in the abdomen—but this guy went down and stayed down. The third body was a tall, robust black man; I was sure he'd sink, too, since black people have denser bones than whites, but he surprised me. Like the first guy, this one was a natural floater.

Tyler left the bodies in the water for five months; by then the flesh was completely rotten, and there was little more to be learned. But along the way he'd observed some interesting phenomena. One of the most interesting was this: Adipocere forms roughly two to three inches above and below the waterline, rather than uniformly over the entire body. We assumed it must be related to the availability of both water and oxygen, but we weren't certain. As with almost any good research project, Tyler's study raised as many questions as it answered.

Up until then, the only research on adipocere formation had been limited to small samples of tissue placed in vials of water in a laboratory. Tyler's project was a truly pioneering study of adipocere formation in its natural setting. Tyler kept careful notes and took numerous photographs; in addition, the university's video department came out and shot a good bit of footage of the experiment. The images on the tape were gruesome, but they were so scientifically enlightening that I included them on an instructional videotape I made for law enforcement officers, as part of a UT continuing-education program called the Law Enforcement Satellite Academy of Tennessee—LESAT, for short.

Unfortunately, a Nashville television reporter who had come to give a presentation at LESAT happened to view that particular tape, and she was horrified by what she saw. That's not surprising; even I have a hard time looking at that footage, and I'm exposed to dead and decaying bodies all the time. I have a hard time seeing footage of surgical procedures, too, but that doesn't mean the surgeon has done anything wrong. In hindsight, though, I could only conclude that this TV reporter mentally blacklisted us and then waited for a reason to pounce.

Before long she got it. By this time Tennessee's medical examiners were sending me a steady supply of bodies that had gone unclaimed after death. Some of those unclaimed bodies were homeless men, and—unbeknownst to me—a few of those homeless men also happened to be military veterans.

I served in the Army during the Korean War. I have the highest respect for the men and women who defend our nation, and I would never intentionally do anything disrespectful to any veteran, living or dead.

But none of that made any difference when Nashville's Channel 4 heard that honorably discharged veterans were rotting on the ground at the Body Farm.

My first warning of trouble came when a reporter called to ask for an interview. "Sure," I said, "come on over." That entire fall I was teaching some 300 miles from Knoxville at UT-Martin, another state school in northwest Tennessee. The reporter and her cameraman made the 150-mile drive from Nashville to Martin. As they were setting up the camera and lights, she told me that she'd dug up copies of every story the Knoxville newspapers had ever published about me. When the camera began rolling, though, her questions focused on just one of those dozens of stories: the 1985 protest at the Body Farm by a local group called S.I.C.K.—Solutions to Issues of Concern to Knoxvillians. Her questions about the protest and other opposition continued for forty-five minutes, then the reporter asked if they could film my class. "Of course," I said, so they did. Afterward she grilled me on camera for another forty-five minutes. I was beginning to understand how people feel when they're in the hot seat facing a 60 Minutes reporter.

A few weeks later my friends from Channel 4 followed me to a guest lecture, camera rolling. I felt as if I were being stalked, and I didn't know why. From the hostile tone of that ninety-minute interview in Martin, I began to fear that they had some hidden agenda, and that concerned me. So when they asked to film at the Body Farm, I told them no.

A few more weeks went by, and one day I got a call from the campus police: Could I please come out to the research facility? When I got there, they were holding the cameraman from Channel 4, who had driven his vehicle up to the facility's wooden gate, set his tripod and camera on top, and begun shooting footage of everything he could see inside the fence.

I was furious. When the TV station first contacted me, I'd bent over backward to be open, honest, accommodating, and fair. If they had done the same, I'd have been happy to continue cooperating, but now I felt betrayed; by this time I'd decided they were on a witch-hunt of some sort. The cameraman called his boss at Channel 4; the station called its lawyer; the TV lawyer called a UT lawyer.

A couple weeks after the guerrilla filming incident, Channel 4 finally aired its report. A four-part series they called *Last Rights*, the story decried what it portrayed as the mistreatment of deceased veterans at the Body Farm. Some of the footage was what they'd shot over the top of our nine-foot wooden fence, but most of it was from the LESAT education video—specifically, the graphic footage from Tyler O'Brien's study of adipocere formation on bodies in water.

To me, the series seemed distorted and lurid, but maybe the TV people thought it was an important blow for dignity and decency; it probably didn't hurt their ratings, either. Whatever their intentions, the story had a powerful impact. For days after it aired, angry veterans, indignant relatives, and irate citizens called me constantly; other calls came from university officials, alarmed by the negative publicity. In retrospect, I suppose something like this was inevitable. For years we'd been conducting research that required us to sidestep society's customary treatment of the dead; for years we'd received modest but positive press when our work helped solve crimes; and recently we'd been thrust into the national limelight by the publication of a best-selling murder mystery. We were a hot topic, and maybe somebody, somewhere, decided we needed to be taken down a peg or two.

I hoped the trouble would die down swiftly, but those hopes were soon shattered. As it turned out, the initial furor proved to be the calm before the storm, because Tennessee's commissioner of veterans' affairs joined the fray. He persuaded several members of the state legislature to sponsor a bill that would have eliminated our research with unclaimed bodies from medical examiners. Given that those bodies accounted for a sizable percentage of our research subjects, the effect would have been crippling.

I was stunned that matters had reached such a crisis. This was the only scientific facility of its kind in the world. In our first few years of research, we'd published pioneering data on the processes and timing of human decomposition, and that basic data was used all over the world. That data had helped police and prosecutors put dozens of murderers behind bars. I myself had testified as an expert witness in dozens of murder trials and helped send more than a few killers to prison. My former graduate students had become scientists whose research at the Body Farm was beginning to establish them as leading experts in their own right. And we'd only begun to scratch the surface. There were so many more variables to study, so many more techniques to develop and refine. . . .

I knew I couldn't fight this battle alone, but I didn't know who could help me. I'd fought scientific battles before, but never legislative ones. If we lost this fight, the Body Farm would go down in scientific history as a bold but doomed experiment.

Then I remembered the prosecutors. They could be the key. There were thirty-one district attorneys in Tennessee, and not only were they law enforcement officials, they were also elected officials: voted into office, and kept in office, because of their commitment to fighting crime. I'd helped a number of the district attorneys directly; in fact, I'd even helped put away a man who had killed an assistant DA in Knoxville a few years before.

I took out my directory of Tennessee law enforcement officials and I began dialing. I told them my side of the veterans' story, I sent a brief history of the research facility, and I explained what it would mean, not just to me but to police and prosecutors, if the legislature curtailed our research at the Body Farm.

Three months after Channel 4 aired *Last Rights*, the anti–Body Farm bill came up for a vote in a key Senate committee. Two of the bill's sponsors served on that committee, so the situation looked grim. But then another senator asked to comment on the bill, and he spoke against it passionately. The bill would effectively close down the Body Farm, he argued, and that would hinder the efforts of law enforcement. "The concerns for the remains of the deceased," he said, "should have to yield to the need to apprehend criminals." The committee voted 5–4 to shelve the bill. We'd avoided catastrophe by the narrowest possible margin.

Sometime later, I happened to be at a meeting where the governor of Tennessee was present. The governor took me aside afterward and, leaning close to my ear, said quietly, "Apparently my commissioner of veterans' affairs doesn't have enough work to do." I took that as a sign that the uproar over the Body Farm was over—for the moment, at least, and for good, I hoped.

CHAPTER 16

The Backyard Barbecue

B ACKYARD BARBECUES are popular in Tennessee
during the summer. I've been to hundreds of them.
One of them was a humdinger.

On July 21, 1997, a TBI agent named Dennis Daniels
called me from a rural area in Union County, Tennessee,
about forty miles north of Knoxville, and asked me to come
take a look at some bones he suspected were human.
Daniels—along with two Union County Sheriff's Department investigators, David Tripp and Larry Dykes—was at
the house of a twenty-one-year-old man named Matt Rogers.

I collared two graduate students who were part of my
forensic response teams, Joanne Bennett and Lauren Rock-
hold, and headed to Union County. We'd had twenty-two
forensic cases so far in 1997; this, then, would be case
97-23. We met a sheriff's deputy at the county courthouse
in Maynardville, then followed him out into the country.

Serious country. The road wound through woods, hardscrabble farms, run-down houses, and rusting trailers; we wound up somewhere around a ramshackle hamlet called Jim Town.

The Rogers house was a small wooden structure; it was painted, or had been, once upon a time, but most of the paint had long since peeled off, leaving the boards to weather to a silvery gray. The officers led me around to the side of the house and behind a toolshed. I knew right away what they wanted me to look at, even before they pointed it out: a rusty fifty-five-gallon oil drum, its sides pierced with large bullet holes. It's what country folks call a "burn barrel"; put a smokestack on it and move it to the city, and it would be promoted to "incinerator." What had caught my eye was the end of a big bone sticking up out of the top of the barrel.

"Matt says they're animal bones," Agent Daniels told me. "A dead goat his dogs drug up into the yard." It was clear the TBI agent didn't believe Matt's story.

Daniels had good reason to be suspicious. Matt's twenty-seven-year-old wife, Patty, had been reported missing eleven days before. Adding fuel to the fire of suspicion was the fact that Patty's disappearance had been reported not by Matt but by Patty's best friend, Angie, who had last seen Patty on July 7 at a cookout. At the cookout Patty had told Angie that she planned to leave Matt the next day. But Angie wasn't the only one Patty told, and that's when the plot began to thicken, like something right out of a soap opera. Patty, it seems, was having an affair with Angie's brother, Michael. That night at the cookout, Patty and Michael told Matt about the affair and said they wanted to be together on the morrow. Patty and Matt left the cookout engaged in a bitter argument.

Angie didn't hear from Patty for two days, which concerned her, given how close they were and what Patty had told her. Then Matt called, and Angie got truly scared: He asked if she'd seen Patty. She'd stormed out of the house at 2:00 A.M. the night of the cookout, he said, and he hadn't seen her since.

The next day Angie went to the sheriff's office to report Patty missing. She'd tried to persuade Matt to file the report, but he'd refused; he'd also asked her to let him know if she contacted the sheriff, so he could straighten up the house before anybody came over to talk to him. Angie did not tell Matt she'd filed the report, and when deputy Larry Dykes went out to the Rogers house, he noticed that Patty's purse, car keys, and cigarettes were sitting on the counter. It struck him as odd that a woman would leave home for three days without those things, not to mention her child.

Patty stayed missing; her daughter went to stay with Matt's parents. On July 21 the missing-person report was turned over to Detective David Tripp. The more Tripp learned, the more certain he became that Patty hadn't simply walked out on her husband and child. It had now been two weeks since anyone had seen Patty. Detective Tripp and Deputy Dykes returned to question Matt again; this time, they brought along TBI Agent Daniels. They also brought cadaver dogs.

Matt Rogers stuck by his story. When Tripp and Daniels asked permission to search his property, he consented. As the cadaver-dog handlers fanned out across several acres, Matt sat down on a rock in the yard to watch the search.

Agent Daniels was drawn to the underside of the house. The house sat several feet off the ground, supported at the corners and several other places, but there was no enclosed foundation or crawl space. Daniels got a flashlight from his car and began peering into the darkness under the floor.

Tripp, meanwhile, noticed a trash pit and the barrel in the side yard, both showing recent signs of burning. A lifelong country boy himself, he knew that when somebody in the country needed to get rid of something, the tendency was to dump it or burn it. Tripp peered into the barrel and called out to Daniels, "You can call off your cadaver dogs. I believe I've found our girl." It was then that Matt, still sitting on his rock, explained about the dogs and the goat, and it was then that Daniels called and asked if I could bring a team out to Union County.

I could see why they might have doubted Matt's story about the goat bones. *I* sure didn't believe it: after forty years of studying human skeletons, I knew a human femur when I saw one sticking up out of a burn barrel. This particular femur was badly burned—its fractured surface and grayish-white color told me it had been burned for a long time in a hot fire—but it was unmistakably human.

The barrel wasn't the only place where a lot of burning had occurred. A few feet to one side lay a mattress; once upon a time it had *been* a mattress, anyhow. Now it was a debris field of bent and blackened springs interspersed with charred tin cans, batteries, broken dishes, and other household trash. Leaning down for a closer look, I spotted what looked to be tiny fragments of burned bone nestled among the debris. We were going to have our work cut out for us. It was already late afternoon by now; we had about three more hours of daylight in which to excavate and recover bone fragments scattered over a large and complex site.

Joanne and Lauren unloaded our gear from the truck: shovels and trowels for digging; wire-mesh screens for sifting rubble; cameras, calipers, and specimen bags. The rubble was strewn over a fairly big area, about eleven feet long by five or six feet wide. To help us keep track of what we found and where we found it, I used surveyor's flagging tape to divide the area into a grid of twelve equal rectangles.

Joanne worked the grid on one side, and Lauren worked the other. Meanwhile, I excavated the barrel, pausing every so often to check on the women's progress. As they worked their way along the mattress grid, it soon became apparent that the body had been burned on the mattress initially, as the fragments there were arranged roughly in anatomical order. Portions that had resisted burning were then transferred to the barrel for additional burning. Most people don't realize how hard it is to consume a body by fire. It sounds like an easy way to get rid of a murder victim, but it's not.

The burn barrel contained a wealth of skeletal material besides the femur I'd first spotted. The femur (it was the left one), while extensively burned, was still relatively intact. Not so most of the other bones in the barrel: most were gray, brittle shards, which I had to handle carefully to avoid breaking. Laying the barrel on its side, I stuck my head in and carefully sorted through its contents, looking for bone. I found plenty, all of it fragmentary: parts of a scapula, a tibia, other long bones, most of the sacrum, and a number of vertebrae. Some of the vertebrae had fallen to the bottom of the barrel and escaped the fire's worse effects; lightly charred, they still had bits of soft tissue on them. A large piece of a cranium was down in the bottom as well, also not as badly burned as the other bones. Scattered on the ground around the base of the barrel were still more bones: additional long-bone fragments, pieces of the sacrum and sacroiliac joint, fragments from ribs and vertebrae, a toe bone, and two more pieces of the skull.

As I was excavating the barrel, Joanne and Lauren were methodically working their way across the twelve rectangles of the mattress-area grid. First they did a visual scan of the surface, where they found numerous bone fragments. Then, once they'd plucked out every bone they could see, they began to screen all the other ashy material, all the way down to bare earth. Three of the grid's twelve rectangles contained trash but no bones; the other nine yielded bone fragments by the thousands. By the time we finished excavating the scene, darkness was falling. Over the course of three hours we'd filled thirty-two paper evidence bags (each about the size of a lunch sack) with bone fragments.

We headed back to Knoxville. Matt Rogers headed off to the Union County jail, where he was charged with first-degree murder.

SOME MEN will do anything to be rid of their wives. I, on the other hand, would have done anything to hang on to Annette.

It took us completely by surprise. On New Year's Eve of 1996, Annette had noticed a couple of swollen lymph nodes along her collarbone. Bright and early January 2, she was at the doctor's office. They took X rays, and the picture was stunning and grim: lung cancer, already at Stage Four. A round of radiation, and the tumor went away.

Just five months later, though, Annette went away too. She awoke one morning struggling to breathe. I called an ambulance. On the way to the hospital, her heart stopped. They revived her; it stopped again. The cancer had come roaring back with a vengeance. Even as the ambulance came racing up to the ER entrance, Annette was dying. By the time I got there—a minute or two behind the ambulance, no more—she was gone.

All my life I'd been a believing Christian. I wasn't without doubts—what thinking person ever is?—but still I had trusted in the existence of a loving God. I'd grown up in the church; I'd taught Sunday school for years; I'd taken youth groups to Mexico for summer mission projects. But that instant in the ER—the instant Annette died—I seemed to feel my religious faith die, too.

As I thought more about it in the bleak days and weeks that followed, I decided the Bible had gotten it exactly backward. Maybe God hadn't created us in His image; maybe we'd created God in *our* image. A Greek philosopher had reached the same conclusion some 2,500 years ago: "The Ethiopians say that their gods are snubnosed and black," wrote Xenophanes, "the Thracians that theirs have light blue eyes and red hair. . . . If cattle and horses or lions had hands, or were able to draw with their hands, and do the works that men can do, horses would draw the forms of the gods like horses, and cattle like cattle, and they would make their bodies such as they each had themselves."*

A loving father: That was the picture of God I'd drawn with my heart, if not with my hands. That's what I'd wanted and needed God to be, ever since that shot rang out in my daddy's office sixty-five years before. But could an all-powerful, all-loving Heavenly Father have allowed these two fine women of mine to die of cancer? Ann had been a nutritionist; besides eating a healthy diet herself, she taught thousands of others to do so, too; yet cancer had struck her digestive tract. Annette, who died of lung cancer, had never smoked a day in her life; her only medical sin was to spend three decades married to a heavy smoker.

* From *The Presocratic Philosophers*, by G. S. Kirk, J. E. Raven, and M. Schofield; © Cambridge University Press, 1988.

Maybe it all boiled down to mere chemistry and genetics: Ann and Annette simply didn't possess enough physiological or genetic resistance to the carcinogens with which the world is filled. Some people do; these two women didn't. Perhaps that was the cold, objective reason they died.

Ann's death had been slow and draining, and I'd begun dealing with it even before it was over. Annette's was swift and crushing, and it came only two months after the death of my mother, who had been very close to me all my life. The weight of grief was staggering. I dreaded setting foot in my empty house. Without warning I would find myself sobbing, unable to stop. Those months were some of the darkest of my life.

All I had left to live for was my work. Cases like this one: a case in which a man was suspected of killing, dismembering, and burning his own wife. The world seemed full of wrong.

THE NEXT DAY, down in the bone lab in the basement of the stadium, we began fitting the pieces of bone together, like some charred jigsaw puzzle. I hoped we'd be able to piece together not just the skeleton but the story of this person's death—probably the story of Patty Rogers's death.

I already knew that the story, like the skeleton, would be fragmentary at best. At the scene we'd recovered pieces of virtually every bone in the body, with one notable exception: apart from a bit of cheekbone, all the bones of the face were missing, and so were the teeth. Teeth are durable—they often survive even commercial cremation fairly well—so their absence, plus the lack of face bones, told me that those parts of the skull had been carefully removed in an effort to make identification of the victim impossible. I wasn't ready to concede that it would be impossible, but it sure wasn't going to be easy.

As in every case, we began by trying to determine sex, age, race, and stature. Lacking the racially distinctive structures of the face, and possessing not so much as a single completely intact long bone—lacking a single intact bone of *any* sort, for that matter—I knew we wouldn't be able to determine either race or stature. Sex and age, on the other hand, we could probably figure out from what we had.

Luckily, one of the hipbone fragments included the sciatic notch. The sciatic notch—the gap through which the sciatic nerve passes when it emerges from the spine and runs down the leg—is markedly wider in females, because the hipbone above it flares out more widely. (The sciatic notch is to the hipbone what the notch beside a long, pendulous earlobe is to the side of the head.) In adult males the sciatic notch has just enough room to accommodate the end of your finger; in adult females there's two to three times that much room. The sciatic notch in this case, case 97-23, was wide, telling us unequivocally that this was a woman's body. One question down, one to go: How old a woman was she?

Analyzing the structure and texture of the pubic bone is often the best way to pin down an age estimate, but in this case those features had been destroyed by the fire. We'd have to look elsewhere for age markers. Fortunately, even though the bones were fractured and fragmented, their epiphyses—the junctions where the ends of the bones fuse to the shafts—remained relatively intact, and the epiphyses can reveal quite a bit about age. Take the femur I saw sticking out of Matt Rogers's burn barrel, for instance. Odd though it seems, as late as age fifteen, that femur had actually consisted of five separate pieces of bone, held together at the epiphyses by cartilage.

Most prominent of the five pieces of an immature femur is the main shaft. Adjoining the shaft's upper end, at the proximal epiphysis, is the rounded femoral head: the ball that fits into the hipbone's acetabulum, or socket. It was the femoral head that first caught my eye in Matt Rogers's burn barrel the previous day. Below the femoral head is the greater trochanter, a prominent, bony bump on the lateral, or outer, part of the upper thigh, right where the leg hinges into the torso. Directly opposite the greater trochanter, on the medial side of the shaft, is the lesser trochanter, a much smaller bump. Finally, down at the distal end, are condyles, forming the femur's half of the knee joint.

The epiphyses can narrow the possible range of a victim's age. They ossify, or turn from cartilage to bone, at different ages. The last of the femur's epiphyses to fuse is the distal one, just above the knee. In some cases, that distal epiphysis doesn't fully ossify until age twenty-two. Since our burned woman's distal epiphyses had completely ossified, she must have been at least twenty-two.

Was there anything else that could narrow down her age? Luckily, even though the pubic symphysis was badly damaged, other age markers on the hipbones had survived the fire. The auricular surface of the ilium (the surface of the hip's broad, ear-shaped upper part) was fine-grained in texture; that, plus the well-defined crest where the ilium fused with the sacrum, told me that she was probably somewhere between twenty-five and thirty-five. So far, at least, I hadn't found anything that would indicate that this was *not* Patty Rogers, a twenty-seven-year-old white female.

Right from the outset we had all figured that these were probably Patty's remains, but over the years I've learned that assumptions can cloud your thinking, leading to scientific error and personal embarrassment. I learned that lesson the hard way in the Colonel Shy case, when I misgauged the Confederate officer's time since death by almost 113 years—my personal record for inaccuracy, by the way. I've also had several cases where the identity of the body turned out to be quite a surprise to homicide investigators. Over in Morgan County, a prominent local contractor disappeared from the town of Wartburg. For years afterward, every time somebody's bones turned up, the police assumed they'd finally found him. They were particularly surprised the time I informed them that their latest find wasn't their middle-aged male contractor at all but an eighteen-year-old female.

So as I began to inspect the shattered bones of 97-23 for some clue, I tried to keep an open mind. It was hard to keep pessimism from creeping in, though. Not a single bone was intact; much of the skull was missing; and everything was burned to a crisp. Correction: *almost* everything. A few vertebrae that were nestled in the bottom of the barrel had emerged largely unscathed, and so had a chunk of parietal bone, from the upper-right part of the skull. Like the other bones we'd collected, the parietal was fractured, but unlike the other shattered bones, the fracture lines of the parietal were unburned. It hadn't been the heat of the fire, or the internal pressure as cranial fluids vaporized, that caused this fracture. Something else—some powerful external force—had shattered the skull around the time of death.

Looking at the other pieces of skull, I spotted what appeared to be telltale traces of that powerful force. The inner surfaces of three different pieces of the skull—the left parietal and two fragments of the occipital, from the base of the skull—bore traces of a grayish-black material, possibly metallic. I had a hunch what it was, and an X ray confirmed that hunch. The material showed up on the negative X-ray films as pure white. That's because it was radiographically opaque: it was lead spatter, from a bullet. Our victim, 97-23, had been shot in the head before her body was burned.

But could we prove that 97-23 was who we thought she was—Matt Rogers's missing wife, Patty? In the absence of facial features or teeth, the only way to make a positive identification would be a DNA test. DNA testing had become widely available about five years before, in the wake of the Gulf War of 1990 to 1991. In this case, though, genetic testing might or might not work: DNA is destroyed by intense heat, and these bones had been subjected to heat intense enough to cremate them, in effect. Our only hope was that the cervical vertebrae or the unburned chunk of the right parietal—the piece that had probably broken away when the bullet smashed into the skull—might yield enough DNA to be compared with samples from Patty's blood relatives. We sent one of the vertebral fragments off to a private forensic laboratory and crossed our fingers as the police requested blood samples from Patty's parents for comparison.

While we awaited the test results, we resumed our examination of the bones. There remained one more crucial question I hoped we could answer: When had she been killed? Joanne was the ideal assistant to help me answer this question. A year before, she had completed her master's degree in anthropology. Her thesis project studied how bone is altered by fire.

Joanne's research looked at bones burned in two kinds of settings. First, she re-created an archaeological setting: she buried prehistoric bones, then built campfires on the ground above them, in order to determine what kinds of changes might have occurred in ancient bones long after they'd been buried—changes that modern archaeologists would need to know how to spot and interpret when excavating ancient sites.

Her second experiment, which was directly relevant to the Rogers case, re-created a realistic forensic setting: Joanne put bones in the crawl space beneath a house, then burned the house to the ground. (Lest anyone think my students are arsonists, let me rephrase that: The house, which had been condemned as unsafe, was burned not by Joanne, but by the fire department, which was kind enough to let Joanne harness the blaze for her research. The fire department's cooperation might have had something to do with the fact that Joanne was dating a firefighter, who is now her husband.)

For her research specimens Joanne used deer bones, which are abundant in Tennessee and are very similar to the bones of humans. She laid some bones on the dirt in the crawl space, buried some bones about an inch beneath the surface, and buried others about two inches deep. Then, with help from a liberal sprinkling of gasoline, the house began to burn.

It burned fast. In just two and a half hours, the wooden house was reduced to smoldering embers. Joanne let it cool overnight, then went back the next day to retrieve her bones and her thermal probes, which measured the peak temperatures to which the bones were exposed. In the crawl space itself, temperatures shot up to about 1,700 degrees Fahrenheit; one inch under the ground, the temperature reached about 1,260 degrees Fahrenheit; and two inches down, it got up to a toasty 1,080 degrees Fahrenheit. The severe heat created numerous cracks in the bones, especially the bones on the surface. Those specimens were riddled with fractures, both longitudinal (lengthwise) and transverse (crosswise, or circumferential).

Joanne's bone specimens for her thesis research were defleshed and dry, but after she got her degree, she conducted additional experiments with "green" bone, fresh bone still covered in flesh. Those experiments suggested that burning fresh bodies creates markedly different fracture patterns: green bone tends to warp when it burns, and its transverse fractures curve or even spiral rather than simply encircling the shaft.

As Joanne and I studied the burned fragments from Matt Rogers's backyard, we compared them to both her experimental specimens and the photos of green bone burned in her later experiments. We were startled to notice that the bones from Matt Rogers's yard weren't warped, and their transverse fractures didn't curve or spiral. Instead, the fracture pattern in case 97-23 bore a striking resemblance to Joanne's thesis samples—that is, to bones that were defleshed and dry when they were burned. Joanne and I both came to an unexpected but inescapable conclusion: The body had decomposed before it was burned. But how had it decomposed so quickly, and where? Those questions nagged at me.

I wrote up our findings, sending copies of the report to TBI Agent Daniels, the sheriff's investigators, and the local district attorney. It wasn't long before I got an answer to my nagging questions. A day after Matt Rogers was arrested, Daniels took a statement from a friend of Matt and Patty Rogers. The friend, named Chris Walker, told Daniels he'd taken a ride in Matt Rogers's car about a week after Patty's disappearance. The car smelled terrible, Walker said—the smell of something dead. When he asked about the smell, Matt told him that Patty's pet turtle had gotten lost in the car and died. The smell was so bad, according to Walker, that he had to hang his head out of the car's window to breathe—an amazing amount of odor from one small turtle.

A few days after his smelly ride in the car, Walker told the TBI agent, he saw the vehicle being towed out of town, in the direction of Knoxville. When he got home, he called a number of wrecker services in Knoxville in an effort to find out where the car had been taken, but he had no luck.

In light of Walker's statement, our findings made perfect sense. The fracture patterns in the bones were exactly what I'd have expected to see, had I known the body was locked in the trunk of a car in the July heat for a week or two. Temperatures in the trunk of a dark car (this one was a blue Buick Regal) can reach well over 100 degrees Fahrenheit for the better part of a summer's day. A week or so of that kind of heat would greatly accelerate decomposition; it would also stink up the car pretty bad, as Chris Walker had noticed.

Walker wasn't the only one who tried to find the missing car. After taking his statement, both the TBI and the Union County sheriff's investigators tried to locate it, but in vain. Rumor had it that the car had been taken to a Knoxville scrap yard, sold for a few dollars, and swiftly shredded. I've always regretted not getting a chance to examine the car; there's no doubt in my mind that my former student Arpad Vass, forensic chemist par excellence, would have been able to take a smear sample of volatile fatty acids and prove that a body had decomposed in the vehicle's trunk.

THE BODY that had probably decomposed in the car—the body that had definitely burned in the yard—was indeed that of Patty Rogers. The bone sample we sent off for testing yielded enough DNA to be reverse-matched to samples from Patty's parents.

At a preliminary hearing, Matt Rogers pleaded not guilty to the charge of first-degree murder in the death of his wife Patty. But on the eve of his trial he took a hard look at the forensic evidence against him. Our report detailed the gunshot wound to the head, the period of decomposition, the removal of the face and teeth, and the otherwise nearly complete skeletal reconstruction. If he stood trial and was found guilty, he could be sentenced to life without parole.

On December 19, 1997, five months after Patty's charred bones were salvaged from a burn barrel and trash pit in the yard of her house, Matt Rogers pleaded guilty to second-degree murder. He was sentenced to twenty-five years in prison.

In life, Patty Rogers had been an unhappy, troubled woman. At one point she was a crack addict, though she claimed to have broken the habit. She'd seriously considered suicide. But in a letter she sent to a friend just two weeks before she disappeared, she wrote that she'd gained some much-needed weight and gotten her teeth fixed. "One day I'm gonna surprise a lot of people," she continued. "I'm gonna make you all proud." Chillingly, the letter contained this request, too: "If God takes me one day, I want you to promise me that you'll see about my kids." I've been told that Patty's daughters are being raised by their father, Patty's first husband, in Florida.

Matt, meanwhile, is serving his time, and I expect it's pretty hard time. He's in Brushy Mountain State Penitentiary, a grim, stone fortress of a prison built a century ago at the base of a forbidding cliff. Brushy Mountain is famous for being escapeproof. Only one prisoner ever came close—James Earl Ray, the man convicted of killing Martin Luther King Jr.—and by the time the bloodhounds and the guards caught up with Ray in the cold, brutal mountains surrounding Brushy, he seemed grateful to be found.

I wouldn't presume to say that Patty Rogers, murdered and burned by her husband, was somehow posthumously grateful to be found. But as a forensic scientist, I was grateful to have had a hand in finding her, a hand in identifying her, a hand in securing for her at least some modest measure of justice. Her story turned out not to be quite as fragmentary as I'd feared it would be. The ending wasn't happy, not by any stretch of the imagination. Grimly satisfying, maybe, and in murder cases, that's about the best ending you ever get.

CHAPTER 17

The Not-So-Accidental Tourist

D EATH AND CRIME know no boundaries, and the bones of the dead speak a universal language, whether they're found in Knoxville, New York, or Old Mexico.

A hundred miles south of San Antonio, Texas, lies Monterrey, Mexico, a city of some three million people. The capital of the Mexican state of Nuevo León, Monterrey, is a bustling industrial center that could easily pass for an American city, except for the abundance of Spanish and the scarcity of pale skin.

On January 17, 1999, my own pale skin—I am a reluctant and nervous air traveler—arrived at Monterrey International Airport. I had traveled to Mexico to meet an insurance investigator named John Gibson and, with any luck, to answer a $7 million question.

Inside the chain-link fence of a police impound lot in

Guadalupe, a suburb on the eastern edge of Monterrey, sat the ruined shell of a Chevy Suburban. Six months before, in July of 1998, the Suburban had burned with enough heat to reduce a man's body to a few handfuls of charred bone fragments.

As with so many other cases, this one began with a phone call from a stymied investigator. Gibson, based in San Antonio, had been hired by a large insurance company, Kemper Life, to look into the death of one of its policyholders. Gibson had already seen the vehicle and what little was left of the person inside. Now he and Kemper Life needed my help in identifying the remains.

Gibson met me at the airport and drove us to the Sheraton Ambassador, a gleaming tower of black glass that would have looked equally at home in Los Angeles or Tucson. Over an early dinner in the hotel restaurant, Gibson filled me in on the details of the case.

The policyholder was an American named Madison Rutherford, a thirty-four-year-old financial adviser from Connecticut. Rutherford and his wife, Rhynie, owned a colonial farmhouse on five acres outside Danbury. They shared their wooded estate with a menagerie of dogs, cats, and chickens. Rhynie was the sole beneficiary of his life insurance.

In my line of work, I'm often reminded of the huge range in the value assigned to different people's lives—and their deaths. Death finds some people so poor, so alone, and so dispossessed that their bodies lie unclaimed in morgues until a county medical examiner or coroner buries them in paupers' graves. Others—blessed with a loving family, social prominence, or hefty insurance—go out in a blaze of grief, glory, and gold. Most of us fall somewhere in the middle. The last time somebody asked me, I couldn't even remember whether I had any life insurance; my wife, Carol, had to remind me that I do. It's a fairly modest amount, though; I'm not worth much dead, and I'm certainly not worth killing.

Madison Rutherford, on the other hand, was worth a fortune dead: a whopping $7 million—$4 million of it through Kemper Life, $3 million more through another company, CNA. Some people would certainly consider him worth killing.

Rutherford and a friend had arrived in Monterrey around July 10, reportedly en route to a dog breeder in Reynosa, a city one hundred miles to the east. There, Rutherford planned to buy an exotic Brazilian dog, a variety of mastiff called a Fila. Rutherford bought a bicycle in Monterrey—a gift, he said, for the dog breeder—and loaded it into the vehicle.

On the night of July 11, Rutherford left his friend at their hotel—the same Sheraton where Gibson and I were now staying—and set out for Reynosa. In the predawn hours of July 12, on the way back into Monterrey, his rented Suburban left the freeway, struck an embankment, and went up in flames. Police and firefighters raced to the scene, but they could do little to fight the intense fire. When it finally subsided, they looked inside and saw nothing—and no one—inside.

Later that morning the police contacted the car rental agency. The agency in turn contacted Rutherford's friend, a retired Connecticut state trooper named Thomas Pietrini. Pietrini asked to accompany the rental agency employee to the impound lot in Guadalupe where the burned Suburban had been taken.

Once there, Pietrini leaned into the passenger compartment, poked around in the charred debris on the floorboards, and emerged with a blackened wristwatch. On the back of the watch was a sooty inscription: *To Madison—Love, Rhynie.* A bit more searching turned up a medical alert bracelet, which warned that the wearer, Madison Rutherford, was allergic to penicillin. Pietrini also found bones—or, more precisely, fragments of incinerated bones. I wondered if there would be anything left in the vehicle for me to find.

ON MONDAY, the day after my arrival, Gibson drove me out to the impound lot in Guadalupe. Over the past thirty years, I've excavated dozens of burned vehicles, but I've never worked one so thoroughly consumed by fire. The glass was gone. The paint—originally dark blue, I think—had blistered off completely, leaving only rusting steel. One corner of the roof had partially melted and collapsed. Inside, virtually nothing but metal had survived: seat frames and coiled springs, the vehicle's own charred skeleton. Seeing the damage confirmed what I had already suspected from Gibson's description of the bones: this had been an incredibly intense fire.

It takes a lot of heat to incinerate a body: After all, by weight, we're mostly water, so getting a body to burn is like starting a fire with soggy wood. But once it finally catches, the human body can burn surprisingly well. One reason is the carbon we contain. The other is the fat we carry.

Several years ago, one of our forensic graduate students studied factors that contribute to cases of "spontaneous combustion"—people whose bodies ignite and burn up. These combustions are actually far from spontaneous, of course. It takes both an ignition source (for example, a smoldering cigarette) and an external fuel source (say, a mattress or sofa) to get the human bonfire started. But in some cases, especially if the victim is grossly obese, the eventual result is a huge, hot, sooty grease fire. I suppose the gruesome moral of this student's research, if research ever *has* a moral, is pretty simple: Watch your weight, and don't smoke in bed. (I halfway do, and I definitely don't.)

Graduate students in UT's anthropology department have actually burned donated cadavers and amputated limbs to gather scientific data about precisely what happens when a body burns. By observing and photographing the processes firsthand, these researchers gather baseline data about the "normal" processes of burning. Armed with such data, we're far better equipped to help police spot abnormal and suspicious patterns. For example, a burning body normally assumes what we call the "pugilistic posture": As the muscles and tendons heat up, they shrink as a result of all that water evaporating, and the hands tighten into fists. The arms flex, too, drawing the fists toward the shoulders in the manner of a prizefighter putting up his guard. The legs bend slightly, and the back arches a bit. It's eerie to see a cadaver actually moving, shifting into a boxer's stance; it seems to be taking one final, desperate stand against the Grim Reaper. Eeriness aside, it's scientifically illuminating. In a real-world forensic investigation, finding a burned body that's *not* in the pugilistic posture could be a clue that the victim was tied up at the time of death, perhaps with the arms bound behind the back.

In this case, though, there was no possibility of finding such clues. For one thing, the remains had already been removed from the Suburban by the Monterrey medical examiner's staff. For another, the heat had been so intense that most of the bones had been reduced to fragments. There was no way to tell whether the arms had been flexed or extended, free or bound, before they crumbled.

Kneeling down beside the ruined vehicle, I leaned in through the driver's door and began sifting through the charred rubble in the floorboard, searching for any remaining bones or teeth. Almost immediately, deep in a layer of rubble, I found a small, gray piece of curved bone. Although it measured only three or four inches square, I recognized it as the top of the cranium. The smooth inner surface had burned away, exposing the spongy bone inside.

Finding the bone within the debris layer answered one question that had been worrying me: Had the body actually burned in the Suburban, or had a set of previously burned bones simply been tossed into the vehicle, either before or during the fire? From the way the other pieces of burned material surrounded it, I could tell that the body had indeed burned there in the Suburban.

But while one important question had just been answered by the cranial fragment, another, equally important question had just been raised: What was the top of the skull doing at the bottom of the rubble heap? And why was it upside down? Theoretically, of course, it was possible that the bone had fallen or been jostled from a higher position, either during the fire or during the subsequent excavation by the medical examiner's staff. However, that explanation didn't fit with the position and condition of the fragment. The inner, concave surface of the cranium had burned away, while the outer surface—the very top of the head—was relatively undamaged. That could mean only one thing: during the fire, the body was head-down on the driver's floorboard.

Next time you're sitting behind the wheel of your car, try an experiment: Invert your position so that your head is down by the gas pedal. Not easy, is it? I know, because I've tried it. Can you envision any scenario in which running off the road and into a ditch—without rolling the vehicle—could possibly shift you into that position? Taphonomically, this case just didn't make sense.

Taphonomy—the arrangement or relative position of the human remains, artifacts, and natural elements like earth, leaves, and insect casings—is one of the most crucial sources of information to a forensic anthropologist at a crime scene. Is there a greasy black stain surrounding the body or skeleton, indicating that the death and decomposition took place in the same spot, or is the ground clean and the vegetation healthy-looking, suggesting that the body was moved or dragged from another location? Are the bones within the clothing, or beside it? Is there a wasp nest in the skull, or a tree seedling growing up through the rib cage? All these things, and many more, are important pieces of the taphonomic puzzle; they can shed a lot of light on when or how someone died.

In Madison Rutherford's case, the taphonomy was topsy-turvy. If Rutherford had run off the highway, crashed in a ditch, and been killed or knocked unconscious by the impact, he would have burned up right where he sat, in the driver's seat. Instead, the body had burned in a head-down position. Even if he wasn't wearing a seat belt, any impact severe enough to cause death or unconsciousness would have triggered the air bag, and that would have limited his movement. The taphonomy was a red flag, a signal that something was fishy here.

After bagging and labeling the cranial fragment, I searched the rest of the vehicle without finding any more bones or teeth. Except for missing that skull fragment, the medical examiner's staff had done a thorough job of excavating the Suburban.

Almost as significant as what we found in the vehicle was what we didn't find. The bicycle Rutherford had bought was gone. On the one hand, its absence could indicate that Rutherford had made it to the dog breeder's and given away the bicycle, as he'd said he planned to do. But on the other hand, there were no dog bones in the Suburban, either. So unless the dog had proved far more adept than the human at escaping the fire, there was a discrepancy between what *should* have been found and what *actually* was found. That was another clue.

So was the fire damage to the vehicle. Aside from the fuel in the gas tank, a vehicle doesn't have all that much flammable material in it: a little carpet, some upholstery, a fabric headliner. But this Suburban had burned with remarkable intensity, so fiercely that firefighters simply couldn't extinguish the blaze. I'm no arson investigator, but I've excavated enough burned vehicles, and talked to enough arson experts, to have picked up some basic knowledge. Judging by the devastating damage to the vehicle, the amount of burning material in the Suburban— the "fuel loading," as arson investigators call it—far exceeded the norm. That suggested that the fire was fed by an accelerant, and lots of it, much of it concentrated at the right rear corner of the vehicle, where the roof had collapsed from the intensity of the heat.

There was one other red flag waving in the breeze above that ravaged Suburban. Supposedly, Rutherford had run off the freeway, gone into a ditch, and hit the embankment so hard that the vehicle caught fire. But there was almost no front-end damage, and Gibson, who had visited the scene of the accident, said the embankment was only slightly scraped or gouged at the point of impact. In short, the "crash" looked like something anyone could have walked away from—or pedaled away from.

But back at the forensic center in downtown Monterrey, there were the bones: clear evidence that someone, presumably Madison Rutherford, had *not* walked away from the four-wheeled inferno.

MONTERREY'S FORENSIC CENTER was a brand-new, sparkling-clean facility, even larger and more impressive than the Regional Forensic Center recently added to the University of Tennessee Medical Center back in Knoxville. When John Gibson and I arrived at the facility, we were met by a small delegation of Monterrey and Mexican government officials. I wasn't quite sure who they all were, as everyone but me was speaking in Spanish, but thanks to Gibson's fluent Spanish I soon found myself in a lab and ready to get down to work. Dr. Jose Garza, a member of the medical examiner's staff, brought me the bones, teeth, and one other item excavated from the Suburban. All that remained of a once-robust man had been scooped up and sealed into a half-dozen or so small plastic bags.

Not surprisingly, most of the bagged bones were calcined, meaning that the organic matter in them had burned completely. These calcined fragments were a lightweight, chalky, crumbly gray—exactly what I'd expect from a really intense fire. And yet, the medical alert bracelet found in the car—stainless steel with a caduceus in red enamel—looked remarkably undamaged. Remarkably unworn, too: its clasp was open.

Any fire hot enough to calcine bone destroys the genetic material, so it's not possible to extract a DNA sample from calcined bone and use it for identification. However, while most of these bones were calcined, not all were. The piece of skull I'd found, for example, would certainly yield enough DNA for a test; so would one or more of the four teeth recovered by the medical examiner. By comparing this DNA with a sample from one or both of Madison Rutherford's parents, who were both still alive, we would be able to tell with almost absolute certainty if these were Rutherford's burned bones. But we had a problem. According to Gibson, Rutherford's parents hadn't provided a sample.

I have three sons, and if one of them was thought to be dead, I'd want to know, and with certainty, whether a body presumed to be his was actually his or not. I can't imagine a parent—any parent—not wanting to know, regardless of the sadness that a positive identification would bring. The lack of DNA comparison samples was yet another red flag. By now this case was raising more red flags than a Chinese military parade.

If we couldn't harness modern DNA testing to confirm the identity of the burned corpse, we'd have to rely on old-fashioned physical anthropology: I'd have to learn the story from the bones. As I began reconstructing the skull, the plot quickly began to thicken. I expected to see cranial sutures that were beginning to fuse, especially on the inner surfaces, where ossification begins. They should still have been readily visible as dark, squiggly lines. Instead, the sutures were almost completely ossified, marked only by slight, barely perceptible ridges of smooth bone, like drywall seams that have been covered with joint compound. Other fragments attested to stout bones with highly developed muscle-attachment points and extensive signs of arthritis.

"You said Rutherford was thirty-four?" I asked Gibson. He nodded.

Four teeth had been recovered from the floorboard of the Suburban: three incisors and a second molar. There were no fillings in any of the teeth. That much, at least, was consistent with Rutherford's dental records. But there were large, unfilled cavities in the two upper incisors—not the sort of thing one expects to see in the teeth of a wealthy financial adviser. The molar was extremely worn, almost like the teeth, which I'd seen in prehistoric graves, belonging to people whose lifelong diet of stone-ground grain had steadily ground down their teeth as well. The incisors bore two other striking features. They were shovel-shaped, square and flat, with a U-shaped rim on their inner side; and their worn edges indicated a classic type of bite.

I called Gibson over and showed him the teeth. "You see that wear pattern? That's called 'occlusal wear,' " I said. "It's caused by the teeth clacking and rubbing against each other. In this case, the edges of these upper teeth lined up almost exactly with the edges of the lower teeth; that's known as an edge-to-edge bite. People of European descent don't have that type of bite."

"Who does?" he asked.

"People of Mongolian descent: Asians. Eskimos. Native Americans."

Gibson stared at me. "So what you're telling me here is . . . ?"

The puzzle pieces—the worn teeth, the invisible sutures—had all fallen into place, and the picture it showed me was not Madison Rutherford's. "This isn't a thirty-four-year-old Connecticut stockbroker," I told Gibson. "This is a fifty- or sixty-year-old Mexican laborer."

A lot of money was riding on the identification of these burned bones. The Kemper Life policy had been issued just six months before the "accident," and when Rutherford bought it, he told Kemper he was canceling the CNA coverage. Instead he was doubling up, and then some.

By now it was obvious that Rutherford had neither died in an accident nor been ruthlessly murdered. He had elaborately faked his death. His tragic death was an elaborate hoax, a $7 million scam. On the basis of my findings, Kemper Life refused to pay the $4 million to Rutherford's "widow," Rhynie. In the delicate, formal language of the insurance industry, "the deceased was not the insured."

Rhynie sued Kemper; she also sued CNA, which was likewise balking at paying their $3 million. The forensic evidence was clearly on the side of the insurance companies. On the other side of the case, though, was a woman who had received a death certificate from Mexican authorities; she had cremated and scattered a portion of the remains, and she now lived conspicuously alone. Despite the scientific evidence, there was some risk that a jury might accept Rhynie's version of the case: heartbroken widow is abused by heartless insurance companies. Both companies reached an out-of-court settlement with her, Kemper for a tiny fraction of the policy's value, CNA for a larger but still modest sum.

Meanwhile, Madison Rutherford—the living, breathing Madison Rutherford—had vanished into thin air, even more thoroughly than if he actually had burned to a crisp. And that, it seemed, was that. For a while.

I PUT AWAY my file on the faked death and got back to my real life. Gradually, out of the ashes of grief after Annette's sudden death, happiness had emerged once more. I owe a big debt of gratitude to my youngest son, Jim, for that turn of events. He was visiting from Atlanta one day during the sad months after Annette died, and I had told him how lonely I was. Out of the blue, Jim said (for it was a suggestion, not really a question), "Why don't you marry Carol Lee?" It was one of those ideas whose brilliance, once it's voiced, is patently obvious—the kind of idea that makes you say, "Why didn't *I* think of that?"

Carol Lee Hicks and I had grown up together in Virginia. She was younger than I was by nine years, but our town was small and our families were close, so we played together often. In fact, I remember playing with her one July day in 1944 at her grandmother's house—a game of hide-and-seek, followed by some lively chicken-chasing. (In southern Virginia in 1944, you took your entertainment wherever you could find it.) Toward lunchtime, as we were running down the road to Carol's father's flour mill, she complained that her side and her leg hurt. "Oh, we're almost there, don't stop now," I said. Then I looked at her, and something I saw made me say, "Okay, let's sit on the bank here for a minute."

That afternoon Carol began running a fever; by the next day it had escalated into chills. Her doctor had just been reading a journal article about polio and quickly realized that Carol was in the early stages of the disease. By getting her to the hospital in Lynchburg right away, he probably saved her life.

Carol walked into the hospital on her own; three days later, when her fever broke, she was already paralyzed from the waist down. She would spend seven or eight months in the hospital, and she wouldn't walk again until early 1945. And she was one of the lucky ones.

Polio has been virtually forgotten by now, but in the first half of the twentieth century, it was a plague of almost biblical proportions. Tens of thousands of innocent children and young adults were killed, crippled, or paralyzed. Polio, a powerful form of viral meningitis, cut a wide and ruthless swath through an entire generation of Americans.

Carol quickly won the battle with the disease itself, but her struggle to overcome the damage it had done would prove lengthy and excruciating, requiring years of physical therapy and twelve complicated surgeries. In Virginia and Atlanta and Warm Springs, Georgia—where President Franklin D. Roosevelt had established a medical institute to help fellow polio victims—teams of doctors labored over Carol, transplanting healthy muscle onto withered limbs, stretching or cutting shrunken tendons, fusing together unstable anklebones. During my junior and senior years at the University of Virginia, I often visited Carol in the UVa hospital, where she began undergoing reconstructive surgeries at the age of thirteen.

Over the years we stayed in close touch. At sixteen she was a bridesmaid at my wedding to Ann. Carol grew up, married a local boy, and had a son, Jeff. Later, she and her husband and Jeff spent two weeks one summer excavating Indian graves with us in South Dakota. Eventually she and her husband divorced and Carol went to work in an office full of doctors, where her positive attitude and wicked sense of humor kept the practice in high spirits. We saw her every time we went up to Virginia for a visit.

And then Carol began coming down to Tennessee: As my mother's health declined, Carol came down to help take care of her, and when Annette got cancer, Carol came down to help take care of her too. Now I was the one who needed taking care of. And then my son, Jim, bless his heart, posed that brilliant question to me: "Why don't you marry Carol Lee?" And so I did. Life, shared with her, became worth living again.

Carol has been informed that she is not, under any circumstances, allowed to die before I do. She assures me, with a twinkle in her eye, that I'll go first. One way or another, I suspect she's right about that. I just hope she doesn't have some secret $7 million life-insurance policy on me tucked away somewhere.

BOSTON'S NORTH END is a hip, high-tech, creative part of the city, filled with loft apartments, artists, and dot-com companies. In the fall of 2000, one of Boston's hottest Web-design firms was Double Decker Studios, whose clients ranged from Boston's mass-transportation agency to media giant America Online. The company's reputation was soaring, and so was its cash flow.

Thomas Hamilton was helping the budding young company manage its financial growth. He'd joined Double Decker a year or so before as its financial controller; his performance since then had put him in line for a major promotion, to chief financial officer. The job would carry a hefty salary and a hefty set of responsibilities.

Kemper Life had hired a Connecticut private investigator named Frank Rudewicz to try to pick up Rutherford's scent in New England. Meanwhile, a Massachusetts detective, Mike Garrigan, was investigating Thomas Hamilton. Everything checked out except for one odd little detail: Hamilton was driving a car registered to Rhynie Rutherford. When the detectives crossed paths and swapped notes, they found that the affairs of Rutherford and Hamilton intertwined with eerie coincidence. When they traded photos, they saw why: Thomas Hamilton *was* Madison Rutherford. After faking his death in Mexico, Madison had slipped back across the border, returned to New England, and been hired in another finance-related capacity under another name.

That wasn't all they dug up. Thomas Hamilton wasn't the first alias Rutherford had employed. In fact, "Madison Rutherford" was an alias, or had been, for a number of years. The slippery scammer had been born "John Patrick Sankey"; he began using the name Madison Rutherford as early as 1986 to fabricate tax returns, obtain the mortgage on his five-acre estate, and purchase his life-insurance policies. It was only a few months before the trip to Mexico that he legally changed his name from Sankey to Rutherford, and then only after a passport application was rejected. Although he and Rhynie appeared to be living well, they were actually deep in debt: Madison had filed for bankruptcy, and the life-insurance scam was a desperate effort to dig himself out of a very deep hole.

Detective Garrigan found one other helpful tidbit: Madison's new life in Boston included at least two new girlfriends. When she learned this, Rhynie was no longer a grieving widow; now she was an angry, scorned woman.

Alerted by the detectives, the FBI moved swiftly. On the afternoon of November 7, 2000, as "Thomas Hamilton" left his office at Double

Decker Studios, FBI agents swooped in and nabbed him. The U.S. government charged him with wire fraud for faking his death and attempting to defraud the insurance companies. With a mountain of evidence against him, including testimony from an embittered Rhynie, Rutherford pleaded guilty to fraud and received a five-year sentence, the maximum allowed. "This is one of the most serious crimes I've seen in this court," the federal judge told him. "This is a crime that caused a lot of people a lot of pain."

The discovery that he was alive and well in Boston cleared up the mystery of Madison Rutherford's fate and whereabouts. But another tantalizing mystery remains unsolved: Whose body was incinerated in that Chevy Suburban outside Monterrey in the predawn hours of July 12, 1998? One thing is certain: Rutherford didn't just dig up some old skeleton from a handy roadside cemetery—the fracture patterns in the burned bones indicated that the body was fresh when it was burned. So the next question is: Where did Rutherford get a fresh body? Once she began cooperating, Rhynie told government officials that Madison said he'd broken into a cemetery mausoleum and stolen a corpse. If that's really where he got it, I guess it's lucky the crypt didn't contain the remains of a thirty-something Caucasoid male. If it had, he might well have gotten away with it, and "Thomas Hamilton"—$7 million richer—might be living a life of luxury in some lavish Boston penthouse instead of doing time in a federal prison.

CHAPTER 18

The Bloody Beneficiary

W HEN THE OUTCOME of a capital murder trial hinges on forensic anthropology, the pressure is excruciating. On the one hand, there's the risk of helping send an innocent man to the gas chamber; on the other hand, there's the very real possibility that a brutal killer might be set free. That high-stakes dilemma was brought home to me recently when I was called for help by a DA prosecuting one of the most cold-blooded murders I've ever confronted.

The call came in May of 1999 from the district attorney's office in Magnolia, Mississippi, the county seat of Pike County. A young family had been brutally killed in the small neighboring town of Summit. A twenty-six-year-old man and his twenty-three-year-old wife had been stabbed repeatedly, and their young daughter had been strangled and possibly molested. Their bodies, bloody and badly decomposed, were found in a cabin outside of town on December 16, 1993. The prosecutor who called me, an assistant DA named Bill Goodwin, knew that the family had been murdered earlier that fall, but the question was, *how much* earlier? Just how long had they been dead before they were discovered? That was the quarter-million-dollar question.

An accurate time-since-death estimate can make or break a murder case. The Zoo Man case had borne that out in a way I would never forget—and maybe never live down. Three of the four victims had obviously been killed while the suspect, "Zoo Man" Huskey, was at large. But the timing of the fourth death—that of Patricia Johnson, whose body I'd pronounced "too fresh for me" and turned over to the medical examiner—became a matter of hot dispute. If Johnson was killed *after* Huskey was arrested for the murder of Patty Anderson, then clearly Zoo Man had an ironclad alibi for one of those four Cahaba Lane murders, despite what his own confession and Neal Haskell's entomological analysis said to the contrary.

By May of 1999, I had been working forensic cases for more than forty years and conducting decomposition research for nearly half of that time. Since the Body Farm's first research back in 1981—Bill Rodriguez's pioneering entomological study—we'd done dozens of decomp studies, under a wide range of conditions. We hid corpses in the woods. We locked them in the trunks and backseats of cars. We buried them in shallow graves. We submerged them in water. Then we studied and documented everything that happened to them, from the moment of death right up until the time, weeks or months later, when nothing remained but bone. We were building a time-since-death database— the first and only one of its kind in the world—by charting the processes and timetable of human decomposition. My goal for the data was simple: Anytime a real-life murder victim was found, under virtually any circumstances or at any stage of decomposition, I wanted to be able to tell police—with scientific certainty—when that person was killed.

By this time my graduate students and I had tracked the decomposition of more than three hundred corpses at the Body Farm. So when Bill Goodwin telephoned about a case in which time since death was crucial and asked if I could help, I felt pretty confident when I answered, "I believe I can."

But my confidence would be shaken, my credibility would be challenged, and events in the courtroom would surprise even me.

THE ADULT VICTIMS in this case were named Darryl and Annie Perry. Their daughter, just four years old, was named Krystal. The fact that the case was coming to trial nearly six years after the murders had occurred told me that this must be a difficult case.

The police had identified and charged a suspect; that wasn't the problem. Circumstantial evidence linked him with the crime; he even had a clear motive. But no hard, irrefutable evidence linked him to the murders: no smoking gun or contaminated knife, no bloody fingerprints, no eyewitness testimony. What's more, he had a strong alibi for two entire weeks before the bodies were found. That's why time since death would prove crucial at the trial: If the defense could convince the jury that the family was alive anytime during that two-week period, the suspect would go free.

As far as anyone knew, the only witnesses to the killings, besides the killer, were the three dead people. I would have to learn the truth from the Perrys themselves. But how? By the time I got the call, the bodies had long since been buried, and the cabin where they were found had been cleaned up and sold. Photographs and notes were all that remained to tell the story of how this young family was killed and, more to the point, *when* they were killed. And so I asked Goodwin to send me every picture he had of the crime scene, especially detailed photographs of the victims' bodies. As I hung up the phone, I hoped I could find enough forensic evidence in those photos to do my job.

Two days later the prints arrived by UPS and I tore open the envelope. It didn't take long to realize that something didn't add up. And if I noticed it, I could be pretty sure the defendant's lawyer, or at least his own forensic consultants, would notice it too.

Half of the forensic picture was clear and unambiguous. The photos showed the bodies of Darryl, Annie, and Krystal to be grotesquely bloated. It was a familiar sight to me, one I'd seen hundreds of times before. By the time the bodies were found, bacteria were well along the way to liquefying the internal organs, starting in the stomach and intestines. As the bacteria digested the soft tissue, they released gases that inflated the bellies like balloons. Beneath and around the bodies was a dark, greasy stain caused by volatile fatty acids being released during the breakdown of the tissues. The hair was beginning to slough off their heads in the characteristic, unified mass we call the "hair mat."

The photos of Krystal were among the most poignant I've ever seen. Krystal's nudity underscored how young, how small, and how defenseless she was. Her genital region was badly decomposed. It wasn't known whether she had been sexually molested, according to the autopsy report, the soft tissues were too far gone to tell. In any case, the image was certainly one of brutal violation.

The average person would look at such pictures, think *My God, what a horrible scene*, then turn away as quickly as possible. For me it's a completely different experience. Don't get me wrong: I abhor death—I've lost two wives to cancer, and those ordeals have made me hate death and despise funerals. When I'm studying a crime scene, though, I never regard it as a death; to me, it's strictly a case. Everything I see and smell is a source of data, a possible key to discovering the truth. I once worked a case involving a house fire in which several young children had burned to death. It wasn't their charred bodies that upset me; it was the glimpse of a tricycle and a few other toys scattered in the yard outside: reminders of the life that had been snuffed out by the fire.

As I studied the photos from the Perry murder scene, I checked for skin slippage, exposed bone, hair loss, and insect activity to see how long the family had been dead. Like every case, it was a scientific puzzle, and

I began trying to fit all the pieces together. By zooming in on each individual piece of that puzzle, figuratively and literally, I was assembling a chronology of events. At the same time, I was shielding myself from the horror portrayed by the picture as a whole.

During decades of research at the Body Farm, I had learned that the events of decomposition occur in a consistent and highly predictable sequence. It's the same in murders anywhere in the world, any time of the year. It doesn't vary—not the *sequence*, that is. What *does* vary, and dramatically, is the *timing*. And the main variable that affects timing is temperature.

On one level that's just plain common sense, of course: a warm body's going to decompose faster than a cold one. I used to tell my students, "That's why you keep meat in the refrigerator, not in your kitchen cabinet." Higher temperature speeds up the work of bacteria as a body putrefies. It also promotes greater insect activity. Bugs, like people, prefer to picnic in the summertime. But to take things from the level of common sense to the level of scientific precision took us years of research into decomposition rates, and how those rates vary with temperature and humidity. Eventually we derived a mathematical formula that quantified all our observations. That formula, coupled with crime scene weather data, allowed us to calculate time since death no matter how the temperatures varied.

The key was a unit of measure called "accumulated degree days," or ADDs: simply put, the running total of the average daily temperature. For example, ten consecutive 70-degree days in summertime would total 700 ADDs; so would 20 wintertime days averaging 35 degrees apiece. In either season, winter or summer, a body at 700 accumulated degree days would exhibit similar signs of decomposition: bloating, "marbling" (distension and scarlet coloration of the veins), skin slippage, and leaching of volatile fatty acids. In our experiments at the Body Farm, we measured ADDs forward in time from the moment of death, noting what stage of decomp corresponded to a given number of ADDs. In an actual forensic case, we performed that same process in reverse, backtracking through the crime scene weather data until we reached the date when the ADDs corresponded with the actual state of decay of a body discovered at a crime scene.

In this case, the crime scene photos showed me that the Perry bodies were moving into the advanced stage of decomposition, in which bloating subsides and the tissues undergo most of their breakdown and liquefaction. In my best judgment the decomposition of the Perrys' bodies indicated they were at approximately 800 ADDs. The next step was to learn what sort of weather they'd had in Mississippi during the weeks before the bodies were found.

I asked Bill Goodwin to send me Magnolia's temperature readings for the months of November and December. Those numbers indicated that it had been a pretty chilly fall. On eight separate nights between mid-November and mid-December, the temperature had dropped to freezing or below. Backtracking in time and temperature, I concluded that the family had been killed somewhere between twenty-five and thirty-five days before they were found.

But there was one thing that didn't quite fit with that picture: the maggots. The bodies were covered with maggots, the larvae of blowflies. Just as bacteria consume a body from the inside out, blowflies start on the outside and eat their way in. Between the microscopic bugs and the macroscopic ones, nature is extremely efficient at reclaiming us: during a hot Tennessee summer, a fresh body can be reduced to bare bone in as little as two weeks. A swarm of maggots covered the faces of Darryl and Annie. Much of the flesh was gone, exposing their skulls beneath. The maggots were also massed at a number of other locations, which corresponded to the autopsy's findings of knife wounds—and therefore blood.

Blowflies love blood. They can smell it miles away. If there's a lot of blood and the weather's warm, they can converge on a body by the thousands. They feed and they lay eggs, which can hatch into maggots just a few hours later.

Darryl had defensive wounds on his hands as well as the fatal wounds to his chest and abdomen. Annie had eight stab wounds in various parts of her body. All of these showed intense maggot activity. So did Krystal's genitals—just the sort of dark, moist opening the insects like. The rest of her body was not as badly decomposed as her parents', and for two reasons: Being much smaller and slimmer than her parents, she would naturally decompose more slowly, a phenomenon we observed many times in our studies at the Body Farm. And because she was strangled, rather than stabbed, there was no blood, so she was less appealing to the flies and maggots.

Some of the maggots I saw in the crime scene photos were a half-inch long, a stage entomologists call the "third instar"; in plain language, that means they were fully mature and close to metamorphosing into pupae and then adult flies. That told me that the maggots had hatched from eggs laid approximately two weeks before. I knew that because of studies we did at the Body Farm in the 1980s. A Ph.D. student of mine, Bill Rodriguez, spent months studying the order and the timing of insect activity in human corpses.

But no matter how closely I looked—with my naked eye and with a magnifying glass—the one thing I didn't see in the photos was a single empty pupa casing. This complicated things. The state of decomposition indicated to me that the Perrys had been killed in mid-November. But the maggots—and the absence of pupa casings—suggested that the murders had occurred around December 2. And the suspect—the defendant—had an alibi from December 2 onward. The prosecution had its work cut out for it. So did I.

Goodwin had first called me on May 18. Two weeks later I made the ten-hour drive to Mississippi for the trial of the man suspected of murdering the Perry family.

DARRYL, Annie, and Krystal Perry had lived in a New Orleans suburb called Marrero; so did Darryl's mother, Doris Rubenstein, and her husband, Michael, a cabdriver. In the early 1990s, Michael—Mike—had bought a small cabin 120 miles north, in Summit (elevation 431 feet), as a place for quiet weekend getaways. In November of 1993 the Perrys went up to stay there.

On November 5, 1993, Mike drove them to the cabin and dropped them off. The young couple was having marital problems, they told relatives, and needed some privacy to work things out. They would have plenty of privacy in Summit, all right: Besides the main highway bisecting it, the town has only a few paved streets, and the sidewalks get rolled up at sunset. The cabin didn't even have a telephone.

Mike drove back to Summit twice in November to see if they were ready to go home. But both times he found the cabin dark and locked, and he'd forgotten to bring his spare key with him. On his second visit, he reported, a neighbor said the Perrys had gotten into a rusty brown van and driven away with two men who looked suspicious, like drug dealers. Nobody had seen them since. Finally, on December 16, he returned again, this time with a duplicate key. Entering the cabin, he found Darryl and Annie lying dead on the living-room floor and Krystal's body sprawled on a bed.

Mike went to the nearest phone—at a convenience store a quarter-mile down the road—and called the Pike County Sheriff's Department. When a deputy arrived, he found Mike out back, behind the cabin. "They're in there," he told the deputy. "They're dead. Their eyes are gone."

Right after the deputy came a Mississippi Highway Patrol officer named Allen Applewhite, who would become the lead investigator in the case. Applewhite was shocked by what he saw in the cabin. The bodies were badly decomposed, and the stench of rotting flesh was overpowering. The corpses of Darryl and Annie were bloated and soaked with blood. Krystal was lying on her back, naked, her face and genitals already consumed by maggots. Applewhite had two daughters himself. He was haunted by the image of this young girl, slaughtered for no apparent reason.

But it didn't take him long to find a possible reason—and a shocking suspect. Twenty-four hours after his 911 call to police, Michael Ruben-stein filed a life-insurance claim for a quarter of a million dollars. The person insured was Krystal, Mike Rubenstein's four-year-old grand-daughter.

When he learned of the policy, Applewhite wasted no time getting a copy of it. Mike and Doris had taken out the $250,000 policy in Sep-tember of 1991, when Krystal was two years old. As he scanned the fine print on the policy, Applewhite read something that made his blood run cold. The policy had a two-year waiting period for benefits. Barely three months after the policy's death benefit could be collected, Krystal was dead. As any good detective will tell you, when there's money involved in a crime, you follow the trail of money. That trail, short and straight, led to Michael and Doris Rubenstein.

It seemed unlikely that a woman would be involved in the killing of her own son and granddaughter. But the police had to consider that pos-sibility. What Applewhite learned about Doris Rubenstein didn't fit with the image of a cold-blooded killer. Doris wasn't a particularly ad-mirable specimen of motherly love and grandmotherly nurture. Her main love seemed to be alcohol and pills. Often she seemed woozy, drunk, or drugged—a woman who was incompetent, maybe even pa-thetic, yet probably not a menace to anyone except herself.

But as the state trooper investigated Doris's husband, Michael, a far different picture emerged: a picture of a man who was competent, shrewd, and deadly. Rubenstein had a long history of insurance fraud, including suspicious fire-insurance claims, staged automobile accidents, and faked injuries involving a large cast of characters. One chilling case years before unfolded in front of a twelve-year-old boy named Darryl Perry, the son of Rubenstein's girlfriend at the time, Doris Perry.

The year was 1979. Rubenstein had just taken on a new business partner named Harold Connor. The two men first met when Rubenstein contacted the local unemployment office, asking for the names of job-seekers who might like to help him produce and distribute a tabloid list-ing local television schedules. Because he would teach Connor the ropes—and because he was taking a chance by hiring an inexperienced partner—he demanded that Connor take out a life-insurance policy naming Rubenstein as the beneficiary. The value placed on Connor's life was $240,000.

The policy was issued in August of 1979. Three months later Rubenstein invited Connor on a deer-hunting trip. Connor declined: He had never been hunting before, and had even told relatives he hated the idea of killing animals. But Rubenstein insisted. To keep peace with his new partner, Connor finally agreed to go. On a cold November morning, they drove to Evangeline Parish, Louisiana, parked on Lone Pine Road, and hiked into the woods. Also on the hunt were another of Doris's sons, David Perry; a man named Michael Fornier, who had recently been paroled from federal prison; and young Darryl.

Connor returned from his first and only hunting trip in a body bag. The story told by Rubenstein and the others painted a classic picture of a tragic hunting accident: As Fornier clambered over a fallen log, his 12-gauge shotgun slipped from his grasp. When the butt of the gun hit the ground, it discharged. Connor, who was directly in front of Fornier, was hit squarely in the back. The blast ripped through his chest and shredded his heart.

Rubenstein told the story to game wardens and then to the police; then he told it to a claims representative for Mutual of New York, which had issued the $240,000 policy. But the insurance company delivered some bad news to Rubenstein: the death benefit wasn't in effect yet. Like many life-insurance policies, this one had a two-year waiting period. Connor's death had jumped the gun, so to speak, by twenty-one months.

Rubenstein responded by suing Mutual of New York, claiming he had not been informed of the waiting period. When the lawsuit came to trial, the insurance company put an expert witness on the stand, a Texas forensic pathologist named Dr. Ronald Singer, who was a specialist in ballistics. Pointing to the angle of the entry and exit wounds, Dr. Singer said there was no way the shotgun could have fired when the butt struck the ground. According to Singer, the gun had to have been level, at shoulder height, to have produced the fatal wound. In other words, the gun wasn't accidentally dropped. It was carefully aimed, cocked, and fired.

When Officer Applewhite uncovered the story about Connor's death and the life-insurance policy, he was struck by the similarities to Krystal Perry's death. He was also struck by one key difference: In Krystal's case, the gruesome death occurred—and the insurance claim was filed—immediately *after* the two-year waiting period had ended. To Applewhite, it appeared that Rubenstein had learned from his mistake in 1979 and was careful to do things right on his second murder. Remarkably, my forensic report indicated that the murders had occurred sometime around November 15, a date that coincided almost exactly with one of his admitted visits to the cabin.

Applewhite spent a year building a case against Rubenstein. But when he took his findings to the Pike County district attorney and urged that Rubenstein be arrested, he didn't get the response he'd hoped for. The DA, Dun Lampton, told Applewhite that he would need hard evidence to prosecute the Perry case, but the only evidence Applewhite had was circumstantial. Granted, the quarter-million dollars appeared to offer a strong motive. Clearly, Rubenstein had a history of shady business deals, fraudulent insurance claims, and probably even murder. And Rubenstein certainly had plenty of opportunity to kill the Perrys: He owned the cabin, after all, and had personally driven the family up there. He even admitted he'd gone back to the cabin on two later occasions. But there was no irrefutable proof of Rubenstein's guilt.

Applewhite was stunned and frustrated. When he told Darryl's biological father, Mack Perry, that no charges would be filed against Rubenstein, Mack wept. But Applewhite promised not to let the matter drop. He continued to follow Rubenstein's trail of insurance fraud and other scams, and the mountain of evidence continued to grow. In September of 1995, he was stunned to learn that yet another person in Rubenstein's life had come to grief: after climbing into a car with Rubenstein one Saturday morning, Laron Rosson, a new business partner, vanished without a trace. Just before his disappearence, he had given Rubenstein a truck-load of expensive antiques that had been purchased with bad checks.

In July of 1998, Applewhite finally saw a ray of hope: That month, a Mississippi jury found a man guilty of drowning his four-year-old son, and the verdict was based purely on circumstantial evidence. That man's motive was a $100,000 life-insurance policy. Applewhite went to Lampton's assistant DA, Bill Goodwin—the prosecutor who had just won that case—and pleaded with him: "Bill, we've got a better case than that here. Instead of a hundred thousand dollars, it's two hundred fifty thousand, and instead of one victim, it's three victims."

Two months later the DA's office took Applewhite's evidence to a grand jury. Rubenstein was indicted for the Perry murders and fraud and was extradited from Louisiana to Mississippi. The case went to trial in June of 1999.

The lack of hard evidence wasn't the only problem the prosecutors were facing. One reason time since death was so crucial in this case was that Rubenstein produced a witness, Tanya Rubenstein—a niece, con-veniently enough—who testified that she'd seen Annie Perry alive and well in a bar in New Orleans. That was on December 2, she said—four-teen days before the bodies were found. And Rubenstein had an airtight alibi for the period between December 2 and December 16. If Darryl, Annie, and Krystal were indeed alive on December 2, then Rubenstein could not have killed them. But if forensic science could show that they were already dead by that date, then the credibility of the niece's testimony—and therefore the validity of Rubenstein's alibi—would be destroyed.

But the defense wasn't about to let that happen without a fight. And this particular battle would be waged over the maggots.

Ever since I first studied the crime scene photos, I'd been fretting over the absence of pupa casings. If I'd seen those, I'd have known for certain that the maggot activity had begun well before December 2. But without them, all I could say for sure was that the maggots had been on the bodies for about two weeks. Clearly the cooler temperatures had slowed the activity of the blowflies and maggots: Blowflies are dormant at temperatures below 52 degrees Fahrenheit, and it was well below that for much of the period in dispute. So I was confident that my estimate of twenty-five to thirty-five days was right. But would the jury share my confidence? That was what was worrying me, especially after the defense hammered hard on the fact that I was not an entomologist.

After only a few hours of deliberation, the jury informed the judge they were deadlocked at 11–1 for conviction. The judge declared a mistrial, and the prosecution went back to the drawing board to prepare for the retrial. To strengthen their case, Goodwin and Lampton called in entomological reinforcements: my former student Bill Rodriguez, who by now was considered one of the world's foremost experts on insect activity in human corpses.

THE RETRIAL started on January 21, 2000. A few days later Goodwin called me to the stand. We went over my qualifications and credentials, including the research studies at the Body Farm, and I was accepted once again as an expert in forensic anthropology. Then, just as I had in the first trial, I explained to the new jurors how I had arrived at my time-since-death estimate.

When the defense attorney's turn to cross-examine me arrived, he quickly began trying to undermine my estimate. First, predictably, he brought up the Colonel Shy case, in which I had misjudged the time since death by almost 113 years. That case, I explained, was why I launched our research program at the Body Farm. Then, as I expected, he zeroed in on the maggots and the fact that they were at the two-week stage. I pointed out how the cool temperatures would have slowed their development, but he kept hammering on the fourteen-day number.

There was one other factor that had to be considered, I argued, one that had become clear to me as I had learned more about the crime scene. Yes, the maggots were at the fourteen-day stage. But the bodies were *indoors,* locked inside the cabin. And this particular cabin wasn't some drafty structure of rough logs and mud chinking. This "cabin" was actually built of solid lumber: two-by-fours laid flat and stacked one atop the other. The man who built it had worked at a lumberyard and apparently could get two-by-fours for free, or almost free, so he built the entire cabin of them. Even the interior walls were made of tightly stacked two-by-fours. There just weren't a lot of openings for insects to penetrate.

The apparent discrepancy between the decomp evidence and the insect evidence wasn't necessarily a contradiction at all, I explained. It would have taken a while for the blowflies to detect the scent of death inside—and then even longer to find a way in through the tightly stacked boards. So the two weeks of blowfly activity constituted only a limit—an absolute bare *minimum*—for time since death. The actual TSD was probably far longer, as the other decomposition markers clearly indicated.

My testimony was followed by that of Bill Rodriguez, my former student. Working independently from me, Bill, too, had estimated the time since death at about one month—again, because of how cold the weather was and how inaccessible the bodies were. Between the two of us, Goodwin hoped, the insect problem had been "exterminated." By the time the prosecution rested its case two days later, I was back in Knoxville. Now it was the defense's turn.

Unbeknownst to Bill Goodwin, the defense had lined up a surprise witness: entomologist Neal Haskell, who had recently testified (along with me) for the prosecution in the Zoo Man case in Knoxville and who had returned to the Body Farm in 1998 to expand his comparison of insect activity in human corpses and pig carcasses. Now Neal was testifying on the other side in a murder case. Fair enough: The world of forensic experts is a small one, and sooner or later someone who's worked with you on one case will end up challenging you on another. But I was totally unprepared for what Bill Goodwin called to tell me about the exchange in the judge's chambers that had ensued when he opposed Haskell's appearance as a surprise witness. According to the defense lawyer, not only was Haskell going to support the defense's claim that the deaths had occurred on or after December 2, he was prepared to testify that I had perjured myself—had lied to help the prosecution—in the Zoo Man case.

Scientific differences of opinion are one thing; accusations of perjury are quite another. This was a slap in the face, contradicting everything I stood for, personally and professionally. Four decades earlier, Dr. Wilton Krogman had drilled a fundamental rule of ethics into me: My role in a case was not to serve the prosecutor or the defendant; my role—my *only* role—was to speak for the victim by uncovering the truth. So when Bill Goodwin had first asked me for a time-since-death estimate in the Perry murders, I immediately asked him not to tell me the prosecution's theory or timetable, and he didn't. If I'd thought the Perry family had been killed on or after December 2, I'd have said so and let the chips fall where they would; to hear that Haskell was attacking my integrity made me furious.

But more troubling than my personal and professional indignation was the damage Haskell's testimony might do to the prosecution's case: If the jury believed Haskell's accusation, they might ignore a compelling body of forensic evidence. At the other end of the phone line, Goodwin listened to my indignant tirade, then asked me to fly back to Mississippi to rebut the accusation of perjury. Wild horses couldn't have kept me away.

BACK IN THE COURTROOM in Magnolia, I sat and waited for my chance to defend my good name. As it turned out, when he took the stand Haskell did not accuse me of committing perjury or lying; he merely testified that I'd been wrong about Patricia Johnson's time since death. Perhaps the defense attorney had been bluffing, perhaps Bill Goodwin had misunderstood. Whatever the reason, I was still eager to explain the circumstances: I would recount the events at Cahaba Lane the day I pronounced her body as too fresh for me, and tell how, when pressed by a police officer, I had guesstimated that she had been out there only a day or two. I would stress—as I had in the Huskey trial—the fact that I had not examined or even touched her body but had sent it straight to the medical examiner. For the hundredth time, at least, I regretted making that unfortunate guesstimate, which had dogged me ever since.

Then, as I waited and worried, something astonishing happened. Rubenstein's attorney called the pathologist from the county medical examiner's office to the stand. During her testimony, the pathologist showed enlargements of autopsy photos. These were photos I'd never seen—photos whose existence I hadn't known of until that moment.

And suddenly there it was. In a close-up of Krystal's face and head, nestled in the roots of her hair, I saw it: a tiny brown object about the size and shape of a grain of wild rice. Looking closer, I saw others as well. From my seat in the courtroom, I leaned over the railing and whispered to Bill Goodwin, "You've got to stop this trial. You've got to put me back on the stand."

Goodwin swiftly requested a recess so we could confer. Excitedly, I told him what I'd spotted in the photo: the empty pupa casings I'd been searching for all along—the ones left behind by maggots as they completed their life cycle and metamorphosed into flies. Just as a caterpillar spins a cocoon, from which it emerges as a butterfly, so a maggot secretes a shelter in which it nestles before it sprouts wings. It's ironic: We think caterpillars are cute and butterflies are beautiful, but we think maggots are repulsive and flies are nasty. But to me, maggots and flies have their own kind of beauty. Especially in this case: They were like an answer to prayer, right there in the courtroom.

Those pupa casings proved, scientifically, that blowflies had been feeding and laying eggs on the corpses for more than two weeks. Even if you assumed that they had gotten into that chilly, tightly built cabin and started laying eggs within minutes, that meant that Annie Perry couldn't possibly have been in a New Orleans bar on December 2. Annie, like Darryl and Krystal, was already dead and decomposing in that cabin on December 2. We had the entomological proof after all; the entire forensic picture now made perfect sense.

On February 3, 2000, the jury retired to deliberate. Just five hours later they came back with a verdict. They found Rubenstein guilty of three counts of first-degree murder. For the murders of Darryl and Annie Perry, they imposed life sentences. For the murder of Krystal—the "money child," as Goodwin and Applewhite called her—they sentenced him to die. It seemed fitting, somehow: The jury was convinced that Rubenstein had executed three of his own family, three people who knew and trusted him; now he was the one to be executed.

Every murder is wrong and brutal in some way, but this case was especially shocking, for its calculated ruthlessness and inhumanity. Michael Rubenstein stabbed his own wife's son. He stabbed his daughter-in-law. He strangled a four-year-old child. And he probably killed two business partners. If my expertise can help put away even one vicious specimen like that, then all my years of study and research have been well spent.

During the trial, Carol and I stayed at the same bed-and-breakfast where Darryl Perry's father and stepmother were staying. They were clearly devastated by the loss of Darryl, Annie, and Krystal. One morning after I had left for court, Mr. Perry, a shy and quiet man, approached Carol in the kitchen. Looking down at the floor, he said to Carol, "Please tell your husband thank you for coming down here to help us." When she looked up, tears were streaming down his cheeks.

Doris Rubenstein filed for divorce after Mike's conviction for the murder of her son, daughter-in-law, and granddaughter. I'm not sure if she ever got it, but I know she never lived to enjoy it: recently she was found dead of heart failure.

Rubenstein is currently appealing his death sentence, and the proceedings and pleadings will go on for years. I can't help reflecting that Darryl, Annie, and Krystal never got the chance to plead for their lives. Rubenstein's execution won't bring back the people he killed, but it might protect others from meeting that same fate.

If there's a hero in this case—beyond forensic science, which made the case against Rubenstein—it's Allen Applewhite, the Mississippi highway patrolman who refused to let this case die. He bulldogged this case for years, even when it appeared there was no hope of a trial. He dug up a mountain of evidence against Rubenstein, a man he later told me he came to regard as "pure evil." Allen was so struck by the dark depths of menace and depravity he unearthed in Rubenstein that he still carries a photo of him in his police car, to remind him just how much can be at stake in building a case against a killer. Allen had wept when the first jury deadlocked and the judge declared a mistrial; when the second jury found Rubenstein guilty, he went home and held his own four-year-old daughter tightly in his arms.

CHAPTER 19

Ashes to Ashes

L LOYD HARDEN—"CHIGGER," his family and friends called him—was an East Tennessee farmer. He and a bunch of other Hardens were born and raised on Harden Road in Birchwood, the name given to a handful of farmhouses sprawled across the broad, fertile floor of the Tennessee River Valley halfway along its hundred-mile run from Knoxville to Chattanooga.

Chigger's eight brothers and sisters scattered out across the valley like a handful of windblown seeds, but Chigger stayed put on Harden Road. His life wasn't an easy one. In school, he never made it past seventh grade; when he was seventeen he learned a hard lesson that he carried with him the rest of his life: He and his nineteen-year-old brother argued over a poker hand and Chigger lost, catching a .22-caliber bullet in the chest from a gun in his own brother's hand. He survived, but the bullet was too close to his heart to be safely removed. He kept it inside him for another twenty-seven years, a reminder of the perils of cards, the precariousness of life, and the crucial difference one inch can make when it involves the relationship between a bullet and a heart.

By the spring of 2000, a life of farm labor had made a burly man of Chigger—not just his muscles but his bones, too, which had grown thicker and stronger to bear the loads they carried year in, year out. He must have looked almost comically oversized, planting strawberry seedlings, pinching the tiny plants between strong, stained fingers and a broad thumb. Now, at age forty-four, Chigger wasn't a young man anymore. His back ached, and he had other, deeper hurts too. On the night of April 17, 2000, Chigger took some painkillers. I don't know how many, but it must have been more than a couple. Instead of killing the pain, the pills killed him.

Chigger had once told his siblings he wanted to be cremated, as his older brother who had shot him (accidentally, the family says) a quarter-century before had been. His sister Suzy asked a nearby funeral home to arrange it, and she bought a fancy brass container to hold his ashes. Chigger's girlfriend was pregnant, and Suzy wanted Chigger's child to have the ashes someday.

The family mourned Chigger's passing at a memorial service; afterward his body was taken to a waiting hearse and driven to a crematorium. The funeral service and cremation cost $3,110.59, including almost $800 for a combustible, cloth-covered coffin. A few weeks later a plastic bag of ashy, cremated remains—"cremains," they're called in the funeral industry—was sent back, and a funeral-home employee transferred them to the brass box. Suzy kept them on her mantel for a while, then gave them to Chigger's girlfriend.

Twenty-two months later, his family watched with horror as a macabre story unfolded on national television. Bodies—unburned, decaying human bodies—were being discovered in Noble, Georgia, on the grounds of Tri-State Crematory. Tri-State was where Chigger's body had been sent to be cremated nearly two years before.

The troubles at Tri-State first became public on February 15, 2002. Inspectors from the Environmental Protection Agency, tipped off by a phone call, inspected the Tri-State property and spotted a human skull on the grounds. The EPA inspector called in the cavalry, and soon dozens of sheriff's deputies and Georgia Bureau of Investigation (GBI) agents were swarming over the grounds. Within hours they found dozens of bodies; in the gruesome days that followed, they found hundreds more: 339 in all, buried in shallow pits, stuffed into metal burial vaults, stacked in the surrounding forest like cordwood, even rotting inside broken-down hearses.

The tip to authorities had come, by a roundabout path, from the truck driver who kept Tri-State's propane tanks filled. During the course of a routine delivery, the driver spotted human bodies on the property. But apparently he couldn't conceal his curiosity (or his shock), because the next time he made a delivery, he was told to get off the property and mind his own business.

Tri-State was a family business. Ray and Clara Marsh opened the crematorium in 1982, and it quickly began drawing business from Georgia, Alabama, and Tennessee, the three states whose borders converge about twenty miles to the northwest of Noble. Tri-State consistently charged less than other crematoriums, and its services—unlike that of most competitors—included picking up bodies from the funeral homes it contracted with, then returning to deliver the cremains a day or two later.

In 1996, Ray and Clara turned the business over to their son, Ray Brent. Business remained brisk; by early 2002, Tri-State had cremated some 3,200 bodies. At least, that's what everyone assumed. Then, on February 15, the awful truth began to emerge.

Within hours after their arrival, the EPA inspectors found several dozen bodies in varying stages of decay. The following day, Georgia's governor declared a state of emergency in Walker County, and authorities grimly predicted that the body count could reach the hundreds. In the first of a long series of legal proceedings, Ray Brent Marsh was arrested and charged with five felony counts of "theft by deception," for accepting payment for cremation services he didn't actually perform. By the following Sunday, the body count was approaching one hundred, and Marsh faced additional criminal charges. Hundreds of investigators were converging on Tri-State, ranging from EPA and Georgia Health Department inspectors to county sheriffs, agents from the GBI and FBI, and disaster-management specialists from federal and state agencies.

One little-known emergency response program that falls under the auspices of the U.S. Public Health Service is a grim organization called D-MORT (pronounced "DEE-mort"): Disaster Mortuary Operational Response Team. Staffed by a wide range of volunteer specialists—medical examiners, forensic dentists, search-dog handlers, forensic anthropologists, morticians, and other professionals who deal in one way or another with death—D-MORT teams are summoned to scenes of mass death such as airliner crashes. (My Knoxville Police Department friend Art Bohanan did a research study at the Body Farm for D-MORT several years ago, as part of an effort to develop leakproof body bags. So far that effort still hasn't fully succeeded.)

One of D-MORT's toughest jobs came in April of 1995, when the Murrah Federal Building in Oklahoma City was destroyed by a truck bomb. Three of my graduate students went out to help D-MORT volunteers identify bodies pulled from the building's rubble. A far greater, sadder challenge for D-MORT, though, came in the wake of the terrorist attacks of September 11, 2001. Hundreds of volunteers risked injury to comb the wreckage of New York's World Trade Towers at Ground Zero; other D-MORT members helped locate and identify the dead at the Pentagon.

Five months after 9/11, as they scoured the pine woods behind Tri-State, members of the southeastern regional D-MORT team were stunned by what they were finding. On Sunday, February 17, one of my graduate students, Rick Snow, got a phone call from a D-MORT officer asking him to come to Georgia immediately. Rick, who had signed on as a D-MORT volunteer some months before, possessed some particularly relevant experience: He had recently returned from a stint overseas, working for the United Nations war-crimes tribunal in Bosnia. For eight months in Bosnia, Rick excavated mass graves and helped identify the thousands of civilians murdered in the name of "ethnic cleansing." The politics and the motives weren't the same in Georgia—the only plausible explanations seemed to be some combination of laziness, sloppiness, and penny-pinching on propane costs—but the bodies and the scope of the task were similar to what Rick had experienced in the Balkans.

Rick arrived on Monday, February 18, to help recover and identify bodies. When he set foot behind the fence at Noble, he must have felt himself transported into a scene somewhere between the Balkans and the Twilight Zone. Bodies were scattered throughout the wooded property. Some were buried; some were stuffed into rusting vehicles and steel burial vaults; some were simply tossed beneath the trees and beside junked appliances, their decaying flesh covered only by rotting cardboard, leaves, and pine needles. On the day Rick got there, the body count reached 139; already 29 of those had been identified by distraught relatives. As the only person on-site with mass-burial expertise, Rick assumed a key role in guiding the search-and-recovery work. The task was greatly hindered by the trees and underbrush covering most of the property, so at Rick's suggestion, a crew with chain saws and bulldozers began cutting the trees and clearing the land, all the way down to the red Georgia clay.

The day after Rick joined the effort, the GBI searched Brent Marsh's house, located at the entrance to the crematorium complex, seeking records that could help shed light on the number of bodies that might be hidden on the property, as well as their identities. While searching the house, they spotted still more bodies in the backyard.

Meanwhile, calls from worried people were pouring into funeral homes across the Southeast. Had their loved one been sent to Tri-State?

If so, were these the genuine cremains on the mantel or in the cemetery, or was the dearly departed actually festering on the Tri-State grounds?

By Wednesday, just five days after the story broke, the cost of the investigation had soared to $5 million, and the body count had reached 242. Aided by the chain saws and bulldozers, searchers found nearly 100 more bodies during the next six days. On the twelfth day, the grisly finds ceased.

The final toll was 339 bodies at Tri-State, and immeasurable heartache among the families who knew, or feared, that one of those bodies belonged to a father, a mother, a sibling, a child. About 75 of the 339 bodies were identified within the first two weeks. Most of those were recent, relatively fresh bodies: easy to recognize, hard to look at. But painful as it must have been to identify a loved one among the bodies retrieved from Tri-State, at least those families got swift closure, or the chance to begin seeking it. For hundreds more people, the uncertainty and pain would drag on and on.

Within days after the EPA inspector's discovery of a skull, the lawsuits began—some against Tri-State, others against the funeral homes that had contracted with the crematorium. That's when I started to hear from lawyers.

On February 21, I got an E-mail from William Brown, an attorney from Cleveland, Tennessee, asking me to analyze the cremains that Tri-State had sent back to Chigger Harden's family. Understandably, the family feared that the cremains might not be Chigger's.

Three weeks later, Bill Brown brought me the cremains. Double-bagged in plastic were a few handfuls of dark gray, ashy material. Including the plastic bags, the entire sample weighed 1,650 grams, or 3.6 pounds. That seemed skimpy: the most recent published study on cremation weights gave the average weight of cremains as 2,895 grams for males, 1,840 for females. (Curious about this subject, I began a research study of my own. Several times a week over the next five months, I went to a cooperative crematorium nearby and weighed cremains before they were sent back to families or funeral homes. After weighing fifty sets of cremains from males and fifty from females, I found the males averaged 3,452 grams, or 7.6 pounds, and the females 2,770 grams, or 6.1 pounds.)

As Brown watched, I carefully emptied the bags onto a clean metal tray, then sifted the material through a 4-millimeter wire screen, which would catch all but the smallest pieces. The bags clearly contained fragments of burned human bones: Even though the pieces were small, I could tell from the smooth, curved surface of some of the fragments that they'd come from the head of a femur (thighbone) or a humerus (upper arm bone). There was also a piece of a bone from the hand; a piece from a foot; and small bits of a metatarsal (foot bone), ribs, a femur, and a tibia (lower leg bone).

Much of what the screen caught, though, was nonhuman material. There was one metal staple—not the kind used for stapling papers together but a big, heavy-duty fastener that might have held together a corrugated cardboard carton, of the sort funeral homes use to ship bodies to crematoriums. (Normally when a body is cremated, it's left in the shipping carton and the entire carton is simply slid into the oven; that makes it easier to handle, and it solves the problem of disposing of the carton as biohazardous waste. Afterward a powerful magnet is used to remove steel objects such as the staples.) The screen had also caught some pieces of what appeared to be burned wood and some fragments of black fabric. The fabric surprised me, since cloth burns at just a few hundred degrees Fahrenheit, while a cremation furnace normally runs far hotter, around 1,600 to 1,800 degrees Fahrenheit. Most puzzling of all, though, were numerous marble-size spheres of a fluffy white substance. *Fuzz balls* was the best term I could come up with to describe them. The fuzz balls weighed practically nothing, but they represented a considerable percentage of the sample's volume. Were they accidental contaminants or deliberately added filler? I'd never seen anything like them, and I told Brown so. I offered to get some laboratory tests made at UT; he agreed that might be a good idea, then he thanked me and left.

I got on the phone and called a textile scientist I knew, who offered to look at the fuzz balls. A professor at UT's Forest Products Center agreed to analyze the fragments that appeared to be wood. I arranged to get samples to them.

These tests would pinpoint the nature of the nonhuman fragments that the 4-millimeter wire screen had sifted out. But that left the bulk of the sample, not quite three pounds of powder and fine particles, which had sifted through the mesh. Visually, the material looked darker than the human cremains I'd seen occasionally over the past forty years, but in a court of law, I knew, I'd have to be more precise than that about what this was—or what it *wasn't*.

When Brown had first contacted me, he'd mentioned that there were indications the cremains from Tri-State might include cement powder, because the authorities' search of the facility turned up numerous bags of cement. Cement looks very similar to the ash that results from incinerating and pulverizing human bones, so it seemed possible that the crematorium might resort to sending families bags of cement powder if they didn't have genuine cremains to send. I searched the scientific literature to find if there was an easy test I could do for the presence of cement.

Cement is mostly powdered limestone, or calcium carbonate. One quick test geologists use to tell if a rock is limestone is to squirt a drop or two of hydrochloric acid on the rock. If the liquid fizzes when it hits the rock, they know it's limestone.

I'd obtained a small quantity of dilute hydrochloric acid, which was sealed in a medicine bottle with an eyedropper. Carefully, I sucked up a few drops into the dropper's rubber bulb and squeezed them onto a small mound of powder I'd put on a metal tray. As soon as the drops hit the powder, they fizzed and bubbled. *It looks like this might be cement*, I thought, *or powdered limestone, anyhow.*

I made one last phone call, to Dr. Al Hazari, a UT chemistry professor I'd known and respected for years. Al agreed to obtain a more detailed chemical analysis of the powdery material; at his direction, I rescreened the cremains another five times to be sure it was free of larger pieces and to mix it uniformly. Then I scooped out 42 grams—about an ounce and a half— sealed it in a small vial, and took it over to the chemistry department.

With any luck, we'd be able to tell the Harden family more soon.

IT DIDN'T TAKE LONG to hear back from my colleague in the Forest Products Center. The sample I'd taken him was burned plywood, he said. That was neither surprising nor disturbing: The cardboard cartons in which bodies are generally shipped and cremated have a thin plywood floor, so they can support the weight of the corpse when the carton is picked up. Without it, the carton might buckle or tear, especially if fluids have seeped from the body.

My textile expert's report on the fuzz balls told me they were a synthetic material—probably polypropylene, he said. Polypropylene is an incredibly versatile plastic. Molded or cast as a solid material, it's used to make things ranging from dishwasher-safe food-storage containers to automobile bumpers. Spun into fibers, it's made into outdoor carpeting, floating marine ropes, and tearproof FedEx envelopes.

Polypropylene is light, strong, tough, and versatile, but it's not heat-resistant. Its melting point isn't much above 300 degrees Fahrenheit— less than the temperature at which chocolate-chip cookies bake, let alone the fierce heat required to burn a body. Accidentally or intentionally, the fuzz balls had clearly been added after Chigger Harden's body was cremated.

If, that is, Chigger's body *had* been cremated. Clearly the sample contained fragments of burned human bones. But were they Lloyd Harden's bones or someone else's? If DNA could survive the cremation process, we could answer that question definitively. Unfortunately, cremation, if done right, burns all the organic material in bone. In a process called calcining, bone is reduced to its primary mineral building block, calcium. The carbon-based DNA molecules—like the carbon in a cardboard coffin or a cotton shirt—burn up completely. Chemically, all traces of human life and identity go up in smoke. So the bulk of our sample—that 2.9 pounds of ashy material remaining after the rusted staple and charred fabric and fuzz balls were sifted out—couldn't tell us whether this was Chigger Harden. All it could tell us was whether most of it was, or had been, a human being.

On April 30, I received the results of the chemical analysis. My chemist colleague Hazari had hit upon an ingeniously simple test to suggest whether the material was human. The human body has a fairly consistent chemical composition. At some point during our school years, most of us learn that the body is mostly water—roughly 60 percent by weight. The other 40 percent is divided among a host of other elements, mainly calcium and carbon. (If humans had ingredient labels, like prepackaged foods in the grocery store, our list of ingredients might start out as follows: water, calcium, carbon . . .)

One ingredient that's practically last on the body's list is silicon. On average, the human body contains just 18 grams of it, or about two-thirds of an ounce. If you evaporate all the body's water and burn off all its carbon in a cremation furnace, you're likely to end up with 5 or 6 pounds of cremains, of which silicon represents less than 1 percent by weight.

Hazari had sent my 42-gram sample to a certified commercial lab in Knoxville called Galbraith Laboratories ("Accuracy with Speed—Since 1950"). He could have tested it himself in a chemistry lab at the university, but a certified lab's precision is frequently tested and well documented, and we wanted to be sure the analysis would hold up in court. A Galbraith technician ran the sample through a spectrographic test called "ICP-OES," short for "inductively coupled plasma optical emission spectroscopy." The ICP part of the procedure burns an unknown material in argon gas, glowing brightly at 18,000 degrees Fahrenheit. Then the OES instrument "fingerprints" the sample, essentially, by reading the wavelengths of the light given off as the sample burns. The final step is to compare the sample's optical fingerprint with those of known elements. It's an analytical chemist's version of the way an FBI fingerprint analyst compares a crime scene print with a database of prints from known criminals.

According to Galbraith Laboratories' analysis, the cremains identified by Tri-State as Chigger's were more than 15 percent silicon. Unless he'd been eating a lot of dirt just before he died, that reading was a lot higher than it should have been. It seemed more likely that the cremains contained filler of some sort—concrete, powdered limestone, or even just plain sand.

Whatever it was, it wasn't right. The cremains that came back from Tri-State should have passed three tests, not unlike the three-part oath every courtroom witness has to swear to: that brass box the Hardens got back should have contained Chigger, the whole Chigger, and nothing but Chigger.

What had really happened to Chigger, and to all those other bodies, down in Georgia? On June 20, 2002, I would get another chance to try to figure it out—by way of a firsthand look.

CHATTANOOGA, Tennessee, lies a hundred miles southwest of Knoxville; about twenty miles southeast of Chattanooga, but a world away culturally, lies the unincorporated Georgia community of Noble. It's a name that now seems mighty ironic.

It doesn't take long for U.S. Highway 27 to make its run through Noble. The four lanes are interrupted by one traffic light, two or three gas stations, and a sprinkling of other establishments offering a few essential goods and services: gas and groceries, hardware and hairdos, several varieties of salvation.

If you weren't looking for it, you'd probably never notice Center Point Road, an unstriped ribbon of asphalt turning off Highway 27 to the east. A sign directs the faithful to Center Point Baptist Church ("Where Jesus Is King"), a few hundred yards down the road on the right. To the left is Roy Marsh Lane, followed by Clara Marsh Lane. Just beyond, across the road, is the long, curving driveway leading to Ray Brent Marsh's home and, beyond and slightly downhill, the Tri-State complex.

The house is a small frame structure, maybe a three-bedroom rancher; out front stands an antique Esso gas pump. Just beyond the house is a wooden privacy fence. In this respect, as in many others, it turns out, the Tri-State grounds bear a striking resemblance to the Body Farm. The chief difference is intent: At the Body Farm, we let bodies decompose only because there's no other way to advance this particular frontier of science. It may sound contradictory, but we hold those decomposing bodies in the highest respect for their unique contribution to forensic research and the quest for killers.

Tri-State's fence encloses two large, barnlike buildings, a tiny hut of an office, and a garagelike building with a rusty metal stack jutting up from one end, where the crematorium is located. The larger buildings contain concrete and metal burial vaults; four months before my visit, those vaults had been stuffed with decaying bodies. Now they were empty.

Off to one side of the buildings, at the edge of the woods, I noticed a broken-down hearse resting on flat tires, rusting in the shade. Opening the door, I caught a foul whiff of decomposition; I later learned that a body had been riding in the back for many months, until the property was raided in February. Nearby stood a mobile home, with another junky hearse parked in front; beside the trailer was a commercial-size barbecue grill, which raised some interesting questions—or simply underscored the irony of the crematorium's nonperformance.

The crematorium building contained virtually nothing but the cremation furnace itself, a massive, industrial-looking oven built mainly of blackened firebrick. At the rear of the furnace, the secondary combustion chamber, which burned any organic material not consumed in the main chamber, looked rusted through in several places, as did the flue above it.

Sliding up the furnace door, I peered inside the main chamber with a flashlight. There wasn't a body inside, I was relieved to see, just walls, a ceiling, and a floor of heat-resistant firebrick and concrete, much of it cracking and crumbling. The floor at the base of the furnace was blackened, greasy, and littered with dirt, gravel, and at least one small, unburned human vertebra—a child's—missed by the GBI and the D-MORT team in their sweep of the property.

I was not the only one who came to inspect Tri-State on this hot summer day. Today was designated a "discovery day" for all the plaintiffs filing suit against Tri-State, the Marsh family, and various funeral homes. All the lawyers involved, for both the plaintiffs and the various defendants, brought in their expert witnesses to inspect the facility. Several former students of mine came over to say hello. One of them, Tom Bodkin, works for the Chattanooga medical examiner; another, Tony Falsetti, teaches anthropology at the University of Florida. I also spotted Michael Baden, a prominent forensic pathologist from New York City, accompanied by a New York forensic dentist. The concentration of forensic firepower in Noble was remarkable.

My visit was cut short when Tom Bodkin, from Chattanooga, stooped down in the driveway area and began to point out human bones—unburned human bones—lying in the dirt. A local sheriff's deputy, standing watch over all the lawyers and scientists, radioed headquarters for instructions. Seal off the site, the reply came crackling back. He herded us all off the property, and within minutes a caravan of sheriff's cruisers and black GBI sedans arrived, looking appropriately like some forensic funeral procession. I'd already seen enough of Tri-State and its cremation furnace by this point anyway. I could see what kind of shape the equipment was in, and it sure didn't look like it had been regularly serviced by its manufacturer.

THE GENERAL MOTORS of the cremation industry is a Florida company with the particularly uninformative name Industrial Equipment and Engineering Company, or IEE. In the summer of 2001, some nine months before the Noble story first came to light, I visited IEE's factory in Apopka, a small town outside Orlando.

The IEE Power-Pak is the company's workhorse cremation furnace. Unlike a funeral-home coffin, which is mainly about elegance and show, a cremation furnace is clearly a product of heavy industry, not built for public viewing. With its drop-front door open, the Powerpack looks like a triple-deep, heavy-duty version of the oven in which Hansel and Gretel nearly became gingerbread. The floor is flat, the top is arched, and the entire vault, stretching some eight ominous feet from door to back wall, is lined with firebrick or refractory (heat-resistant) concrete.

Bodies generally arrive by hearse; at most crematoriums, the hearse backs up to a garage door, and the body, in its cardboard shipping box, is pulled onto a gurney, which is wheeled up to the furnace door. It's a simple, one-person operation to slide the box from the gurney into the furnace, then close the door and fire up the gas.

The first step is to switch on a powerful fan, which forces a steady draft of air through the oven—the "primary chamber," it's called—and then out through an exhaust stack. Once the fan is running, the operator sets a timer that controls the length of the burn. The timer also controls a gas valve and a sparking ignition device, much like the ones found in residential gas ranges.

The first burner to ignite is called the "afterburner." Located at the rear of the furnace, it's a small burner that does double duty. First it slowly preheats the interior, to minimize heat stress and cracking in the firebrick. During the cremation itself, it burns any uncombusted gases before they go up the stack.

Once the furnace is preheated, a low-intensity burner, the "ignition burner," kicks on in the roof of the furnace, its flame directed downward. The ignition burner's only job is to incinerate the cardboard carton or the body bag containing the corpse. Cardboard catches fire at around 500 degrees Fahrenheit; engulfed by the flame shooting down onto it, the carton catches fire within seconds.

Several minutes later the cardboard has been reduced to ash, and the cremation of the body itself can begin. Now a more powerful burner, the cremation burner, blasts downward onto the body. In most cases the furnace's temperature stays somewhere between 1,600 and 1,800 degrees Fahrenheit; extremely obese bodies, however, can burn at much higher temperatures, up to 3,000 degrees.

IEE builds its furnaces tough to survive those kinds of conditions. The company also offers annual inspections, cleaning, thermostat calibration, and repairs at the customer's request. Most facilities request at least one inspection and calibration a year. In twenty years, I heard, Tri-State never requested a single inspection or cleaning. Reportedly, the only visit an IEE technician ever made to Tri-State was when the GBI asked the company to come verify or refute Brent Marsh's claim that he'd fallen behind in his work because the furnace was broken. According to the IEE technician, the furnace fired right up.

ON SEPTEMBER 3, 2002, the day after Labor Day, one of Chigger Harden's relatives got a phone call from Greg Ramey, the GBI agent heading the Tri-State investigation. Samples of DNA from the 339 corpses recovered from Tri-State had been analyzed at the Air Force DNA laboratory in Maryland. The lab compared the GBI's samples with known genetic material donated by relatives or obtained from medical offices. Some of Chigger's relatives had given blood samples, but as it turned out, they needn't have: a tissue sample from Chigger's autopsy remained on file at an area hospital.

Agent Ramey called to say that a DNA match indicated that Chigger's body was one of the 339 corpses found on the grounds of the crematorium back in February. Designated by the GBI as body number 218, he had lain decaying in the Georgia woods for nearly two years. Since February he'd been in a cold-storage facility set up by the GBI near Noble. What, Ramey wanted to know, would the family like done with the body now?

The Hardens still wanted the body cremated, in accordance with Chigger's wishes. First, though, they wanted to be absolutely sure it was Chigger. Bill Brown, their attorney, asked me to examine the body, and he arranged to have it delivered to a place where the examination and cremation could be done in swift succession.

One crisp October afternoon I arrived at a small, neat building that houses the East Tennessee Cremation Company, located at the edge of an industrial park near the Knoxville airport. A few minutes later Bill Brown arrived, along with his assistant, Lisa Scoggins, and his son, Andy, who would photograph and videotape the body and my examination, so there would be a visual record for their legal case.

The crematorium's manager, Helen Taylor, ushered me into the garage area, which housed two IEE cremation furnaces, both of them spotless. In front of one was a gurney, and on it was a white body bag. Unzipping it, I found that the body was mostly skeletonized, though bits of tissue remained here and there. Beside the skull, though no longer attached, was the hair mat: long, thick brown hair, just like the shoulder-length brown hair depicted in the photo of Chigger that Lisa had brought.

The body was nude; the clothes had been removed by the GBI and put in a separate plastic bag. Leaf litter and pine needles were scattered through the remains and the clothing, suggesting that the body had been lying outdoors for quite some time. The absence of dirt in the nasal passages and ears told me that the body had never been buried. Here and there, I found small bits of rotting cardboard, as well as a handful of dead dermestid beetles, sometimes called hide beetles or carpet beetles, which like to nibble dried flesh off bones.

The skeleton was largely intact; the lower jaw and the bones of the lower right leg and foot were missing, though, probably carried off by scavengers. I studied the skull. It was large and broad, with a heavy brow ridge and an unusually prominent bump at the base of the skull, the external occipital protuberance. Any student in my osteology class would have had no trouble telling this was a male. The teeth were vertical, rather than jutting forward, so the skull was clearly Caucasoid, and the cranial sutures showed a degree of fusion typical of a man in his forties. Nothing in the skeletal material contradicted the GBI's identification.

The DNA sample had come from a piece of bone taken from the middle of the left femur. Brown asked me to obtain another bone sample, so an independent DNA lab could cross-check the government's results. I unpacked the Stryker autopsy saw I'd brought from the anthropology department and plugged it in.

A Stryker saw is an ingenious tool. It can chew through a femur in a matter of seconds but can also buzz against a child's forearm without even breaking the skin. The secret is that its fine teeth, about the size of those on a hacksaw blade, oscillate back and forth in tiny strokes, just a sixteenth of an inch long. Bearing down on rigid material, such as a corpse's bones or a child's plaster cast, the teeth take rapid little bites. Pressing lightly on the skin, though, the teeth merely wiggle the skin back and forth, maybe producing a tickle and a giggle.

I cut into the femur right beside the notch left by the GBI's Stryker saw; it took less than a minute to notch out a quarter-cylinder about two inches long and an inch wide. I gave this to Brown for shipment to the independent DNA lab. As one final precaution, I bagged and gave him a finger bone, too, in case a third test should someday seem necessary.

Next I opened the plastic bag of clothes tucked down at the foot of the body bag. The body itself wasn't too smelly, but the clothes reeked of decomp and ammonia. Despite their rotting and stained fabric, a pair of blue jeans was easily recognizable. The shirt was also crumbling, but it looked to be red and green plaid. According to Lisa, the Harden family had asked the funeral home to dress Chigger in his favorite outfit, jeans and a plaid shirt.

If we were lucky, we would find one final piece of identifying material accompanying Chigger's body: the bullet his brother had fired into his chest more than twenty-five years earlier. It would be difficult and time consuming to search the remains for it now; I decided my chances of finding it might be better after the cremation, when I could sift through the ashes.

As the sun sank low over the red- and gold-leafed Tennessee hills outside, I folded the white body bag back over the moldy skeleton; with a quick push, the bag slid deep into the furnace. Helen Taylor slid the door upward, latched it in place, and switched on the fan. Seconds later I heard a soft *whump* as the gas ignited.

The next morning dawned foggy and cold. Back inside the garage at the East Tennessee Cremation once more, I felt the heat still radiating from the furnace's masonry. The cremation had taken only a couple of hours, but the body had remained overnight in the furnace so I could examine the burned bones in situ. Sliding the furnace door open, I peered inside the long, dark chamber with a flashlight. The bones inside still clearly outlined a human skeleton. The long bones of the arms and legs were fractured but intact, as was the pelvic structure; the crumbling remnants of a rib cage still sketched the framework of a chest. Most recognizably human of all was the skull. As soon as I touched it, it broke into small pieces.

Using a long-handled broom and a large dustpan, Helen Taylor scooped up the bone fragments and ashes, then spread them on a worktable beneath a vented exhaust hood so I could sift through them. Lurking amid the bone shards and the soft, ashy material were dozens of rusty steel staples; two years before, they had held together the cardboard container in which the body had arrived at Tri-State. Helen handed me a large, heavy magnet and showed me how to drag it through the cremains, trolling for staples.

The weight of the magnet was enough to crush all but the largest pieces of bone; in their lightness and fragility the fragments resembled the airy and brittle meringue cookies made by baking beaten, sweetened egg whites. Tiny blobs of an amorphous, glasslike material were scattered throughout the cremains, possibly left by melted buttons or other artifacts from the clothing that was burned along with the body. As I continued to sift and stir the material, sorting out pieces that were obviously not human cremains, I strained for a glimpse of a bullet or, more precisely, a melted blob of lead, something that might once have been a bullet. I saw nothing that looked even close.

The final stage in the cremation process was to pulverize the remaining pieces of bone. Some of the Tri-State cremains I'd analyzed had contained large bone fragments; news reports had indicated that the Marshes had used a wood chipper or simply a large board to break up large fragments. So I'd done an experiment in cremains processing myself, on another set of cremains I'd received from Bill Brown: I'd put some of the burned bones down in Carol's blender, an old Hamilton Beach model, and switched it on. A terrible clatter and chatter ensued, some of it from the blender, some of it from Carol. (You'd think I'd have learned, after buying two new stoves for Ann, not to use my household appliances for research. Needless to say, the kitchen counter soon acquired a new blender, and the contaminated one was banished to the garage.)

East Tennessee Cremation Services had a much more sophisticated way of pulverizing burned bones: an IEE processor, which looked a lot like a soup kettle grafted on top of a garbage disposal, but cost a whopping $4,000. Helen put the cremains in the kettle and covered the top with a heavy lid, then flipped a switch. The fragments were gone in sixty seconds, reduced to a grainy powder. Then she poured the processed cremains into a plastic bag positioned inside a rectangular plastic case— they fit, but just barely—sealed the bag tightly with a plastic cable tie, and handed the box to me. I now held in my hands what the Harden family believed they had received more than two years before. I placed the container in the backseat of my truck and headed home.

The initial, bogus cremains of Chigger Harden had weighed in at 3.6 pounds, less than half what my measurements of one hundred sets of cremains had shown to be the average weight of cremains from males. The cremains I had with me now, on the other hand, were a testament to a burly farmer's frame: including the weight of the bag (but not the plastic box), they tipped the scales at 8.1 pounds, probably pretty close to what he'd weighed back when he first came into this world. After weighing the cremains, I opened the bag and filled a clean plastic film canister with a sample, then resealed the bag. I sent this sample, like the others, to Galbraith Laboratories.

When the results came back, I was surprised: the cremains contained 5 percent silicon, roughly ten times what I'd expected. Perhaps
all that silicon came from soil clinging to the body or the clothing, or
perhaps some of it was bits of the furnace's concrete lining flaking off.
Another cremains sample, which Galbraith analyzed at the same time,
contained just 0.5 percent silicon, which was much closer to the human
body's normal ratio. As usual, a piece of research had raised as many
questions as it had answered. But the fundamental question had been
answered pretty conclusively by now: We had a positive DNA identification from the GBI and the Air Force; we had an anthropological examination of the skeletal remains, which were consistent with Chigger's
age, race, sex, and hair length and color; we had clothing that matched;
and we had independent corroboration from the commercial DNA lab
that tested the piece of bone I'd cut from the femur with the Stryker saw.

There was one loose end that still nagged at me, one unanswered
question that still kept me from laying this case to rest. I climbed into
my truck and headed for UT. Displaying my TBI badge conspicuously
on the dashboard, I parked in an illegal spot (the only kind I could find)
and walked into the radiology department in the basement of UT's student clinic. Over the years the technicians and physicians there have
been unfailingly gracious and accommodating about my occasional re-
quests to have odd things x-rayed. They seem to find it interesting; they
also seem to appreciate the fact that I don't bring them decomposing
bodies to x-ray: I get those scanned with a portable machine on the
loading dock at UT Medical Center.

From a cardboard box I was carrying, I took out two flat plastic bags,
roughly a foot square, into which I had divided Chigger's cremains.
Spread to a uniform thickness, the cremains formed a square layer about
an inch thick in each bag.

The radiology technician stepped behind her lead shield and opened
the shutter. The first negative she brought me was almost clear, indicating that it was badly underexposed; apparently she'd overcompensated
for the thinness of the sample. Her second exposure was right on the

money: The ground-up bone fragments appeared in many shades of gray; dozens of tiny white toothlike objects dotted the image—metal teeth from the zipper of the bag in which the body had arrived from Georgia and been cremated.

The negative showed one other radiographically opaque object. It was an almost perfect disk, about the size of a penny and twice as thick. I fished it out. The disk was heavy—heavy as lead. I hadn't been able to see it or feel it in the cremains, but it had been there all along. I had found Chigger's bullet.

The Harden family's long period in limbo was over. The finding of the bullet wasn't exactly good news, but they were grateful for it all the same. I've encountered that response time and time again in dealing with the families of the missing and the dead. Uncertainty and dread are almost always harder to bear than the finality of certain loss.

I can't give people back their loved ones. I can't restore their happiness or innocence, can't give back their lives the way they were. But I can give them the truth. Then they will be free to grieve for the dead, and then free to start living again. Truth like that can be a humbling and sacred gift for a scientist to give.

CHAPTER 20

And When I Die

I N MY FIRST FORTY YEARS as a forensic anthropologist, I saw hundreds of corpses and thousands of
skeletons. I scrutinized death from every angle. Every
angle but one, that is. Then one day I found myself flat
on my back on a restaurant floor, staring death straight in
the eye. And death was staring right back at me.

My wife, Carol, and I were driving from Nashville back
to Knoxville. It's about a three-hour drive, and we decided
to stop for lunch about halfway back, in Cookeville. We
pulled off of Interstate 40 and headed for my favorite local restaurant, Logan's Road House, which serves a baked
sweet potato I crave.

I'd gone to Nashville to lecture to a group of organ-
donor professionals. I didn't feel well the night before, and
if I'd had any sense I'd have canceled my talk right then,
but I'd come to Nashville to lecture and, by God, I was

going to lecture. There's a long tradition in the Bass family of a trait we like to call determination. I'm told other people often refer to us as mule-headed.

I gave the group a slide-lecture introduction to forensic anthropology. It starts with the case of a Texas man who set his car on fire and killed himself and moves on to the case of Madison Rutherford, who faked his death in a car fire. I've given this talk dozens of times, but I could barely get through it that morning. Normally in front of a group I come to life: I feel energized, excited; I'm full of stories and jokes. This time, though, I was clearly chugging. Finally, mercifully, it was over. I accepted polite Southern compliments on my lackluster talk, said a few quick good-byes, and hustled Carol to the car, counting big on that baked sweet potato along the way to perk me up. We walked into Logan's and minutes later it arrived, buttery and steaming.

I remember taking about two bites of the potato. Suddenly things began to go black. Pushing my plate to one side, I told Carol, "I'm about to pass out"; with that, my head dropped to the table. I have no recollection of the events that followed; I relate them as they were recounted to me by Carol and others.

Paramedics soon arrived and so did Dr. Sullivan Smith, the county's medical examiner, who had been driving nearby when the 911 call came. Hearing the emergency dispatcher on the police radio in his car, he hotfooted it to Logan's. If he'd arrived a minute later, he might have had occasion to document my death. As it was, he joined the fight to reverse it.

I'd known Dr. Smith for years, ever since his medical residency at UT Medical Center in Knoxville. I consider him one of the state's best medical examiners, and over the years I've lectured half a dozen times or more for his seminars for ER personnel. Amazingly, Dr. Smith recognized me from a glance at the back of my head. (I'm not sure whether that says more about the keenness of his mind or the strangeness of my head.)

"Dr. Bass? Dr. Bass, can you hear me?" he asked, then looked at the paramedic who was still checking for a pulse. The paramedic just shook his head. "Dr. Bass, we're going to have to move you to the floor right now," Smith said, as if I could hear.

They unpacked their portable defibrillator, brought the paddles to my chest, and prepared to give me a jolt of electricity—a final, desperate effort to jump-start my heart. Just at that moment my heart shuddered back to life. Consciousness and vision returned, and I found myself lying on the floor, surrounded by feet—dozens of feet.

"Dr. Bass, can you hear me?" The voice was vaguely familiar; so was the face of the man kneeling over me. ". . . Sullivan Smith," he seemed to be saying.

"Sullivan Smith? Oh, yes, I know him," I murmured weakly. "I've lectured for him."

"No, Dr. Bass, this *is* Sullivan Smith," he said. Finally the fog cleared and I recognized him, grateful to be in such good hands. Another minute or so, Smith said, and they might not have been able to bring me back.

Within hours Smith arranged for an ambulance to transfer me to UT Hospital in Knoxville. The EMT and I chatted for most of the two-hour ride, about subjects ranging from forensic cases to UT football. One thing we did not talk about was my near miss with death.

My cardiologist, John Acker, said my heart muscle itself was fine. The problem was in the electrical system controlling its contractions. Fortunately, the fix was simple: a pacemaker, a sophisticated heart monitor and miniature defibrillator packed into a disk not much bigger than a silver dollar. If my heart was working fine, the pacemaker would do nothing; however, if my heart rate dropped below fifty beats per minute, the pacemaker would kick in.

It was an odd feeling to be in UT Hospital as a patient. I've spent thousands of hours there since moving to Knoxville in 1971: The Knox County Morgue and the Regional Forensic Center are housed in the hospital, and I've examined hundreds of bodies and skeletons there. The fact that I myself now had one foot on the edge of the grave made me all too aware of the proximity of those basement autopsy rooms. A few days later I had surgery to implant the pacemaker.

. . .

ONCE UPON A TIME I believed in an afterlife. I believed in it for fully sixty years after my father shot himself. But then Ann died, and then Annette died, and suddenly nothing I'd grown up believing about God and heaven made any sense to me any longer. We're organisms; we're conceived, we're born, we live, we die, and we decay. But as we decay we feed the world of the living: plants and bugs and bacteria.

People who knew my father—the man I never got a chance to know; the man who eluded me when I was three—say I resemble him in many ways: in my curiosity and intelligence, my friendliness, my kindness; in the way I stick out my tongue slightly when I'm concentrating hard. I'm proud to see these same qualities in my grown sons, and I notice with delight that one of my granddaughters, when she's coloring or practicing the knitting stitches Carol has taught her, sticks out *her* tongue in that distinctive Bass manner. So *something* of us lives on, in some fashion, in those we leave behind us: our genes, our mannerisms, our shared experiences and oral histories.

Is that all that endures? Almost, I think, but not quite. Charlie Snow, who brought me in on my first forensic case—that woman's soggy, burned body we exhumed and identified outside Lexington—is still alive to me in a way when I arrive at a crime scene and begin trying to make sense of what I see and what I smell. So is Wilton Krogman, the Socrates of "bone men": there's a part of me that's forever in a car with him, perpetually commuting to the University of Pennsylvania; in my mind I go over my latest case with him, outlining my conclusions and marshaling my arguments and references to answer any question and refute any objection the great man might voice. I still beam with pride, after all these years, when I spot something Krogman might have overlooked if he had been on the case.

And so it will be, perhaps, with my students. For some of them, I hope, I will always be looking over their shoulder at the shattered skull, the burned bones, the telltale insects; always questioning them, always challenging them, sometimes even inspiring them. There's a part of me that will live on, too, at the Body Farm, my proudest scientific creation. Looking back over the past quarter century, I'm amazed at the wealth of pioneering research that has emerged from such humble beginnings—it began in an abandoned sow barn—and even today the Anthropology Research Facility remains a simple metal shed and a patch of trees and honeysuckle vines, tucked behind a high wooden fence (recently enlarged and rebuilt with help from Patricia Cornwell). That, plus a generation of bright, inquisitive minds eager to unlock the secrets of death. I certainly didn't set out to create something famous there. I just set out to find some answers to questions that were nagging me. As in life, so in science: One thing leads to another, and before you know it, you find yourself someplace you never imagined going.

One question I'm often asked, especially by journalists, is this: "When you die, will your body go to the Body Farm?" Will I practice what I preach; see it through to its logical conclusion? There was a time when I was sure I would. I discussed it with my first wife, Ann, who was also a scientist; she approved heartily. My second wife, Annette—who was my assistant for years and was all too familiar with the facility and its work—said, "Absolutely not." As for Carol, she seems to be leaning toward a more traditional, and—to her way of thinking, at least—more dignified final resting place for Dr. Bass. I'll leave the final call to Carol and the boys. The scientist in me wants to sign the donation papers. But the rest of me can't forget how much I hate flies.

Either way, you'll still find me at the Body Farm when I die. Not anytime soon, though. I don't want to die now. I have too much to do. Books to write. Grandchildren to play with. Killers to catch.

Bones of the Human Skeleton

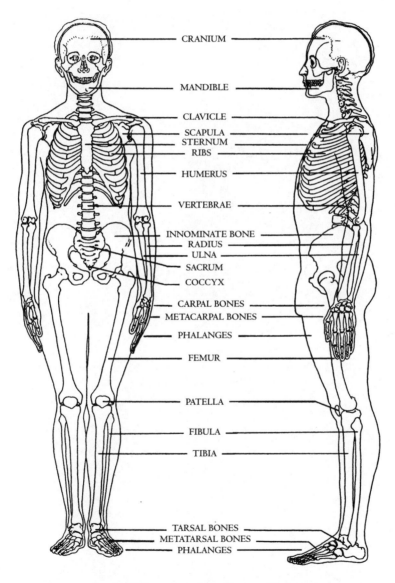

CRANIUM

MANDIBLE

CLAVICLE
SCAPULA
STERNUM
RIBS

HUMERUS

VERTEBRAE

INNOMINATE BONE
RADIUS
ULNA
SACRUM
COCCYX

CARPAL BONES
METACARPAL BONES

PHALANGES

FEMUR

PATELLA

FIBULA

TIBIA

TARSAL BONES
METATARSAL BONES
PHALANGES

General Elements of the Human Skeleton

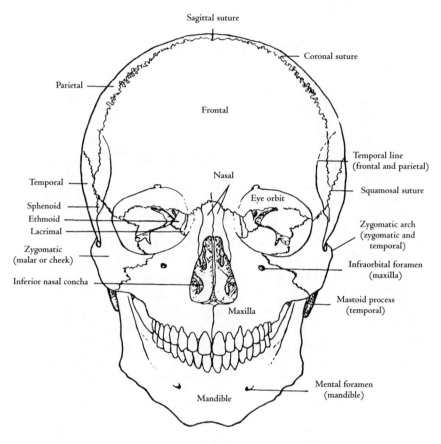

Sagittal suture

Coronal suture

Parietal

Frontal

Temporal line
(frontal and parietal)

Squamosal suture

Temporal

Nasal

Eye orbit

Sphenoid

Ethmoid

Lacrimal

Zygomatic arch
(zygomatic and
temporal)

Zygomatic
(malar or cheek)

Infraorbital foramen
(maxilla)

Inferior nasal concha

Mastoid process
(temporal)

Maxilla

Mental foramen
(mandible)

Mandible

Skull, Frontal View

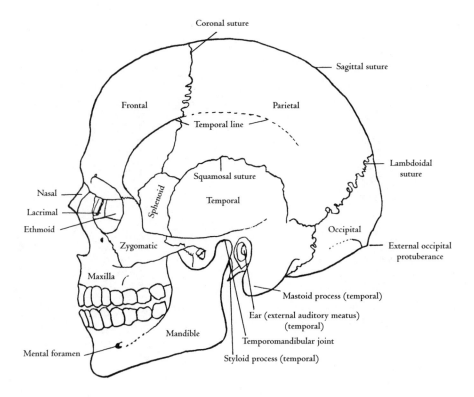

Skull, Lateral View

The illustrations in this appendix are reprinted from *Human Osteology: A Laboratory and Field Manual* (Fourth Ed.), by William M. Bass. © Missouri Archaeological Society, Inc., 1995, and used by kind permission.

APPENDIX II

Glossary of Forensic and Anthropological Terms

accumulated degree day (ADD). The cumulative total of the average daily temperature, measured in either Fahrenheit or Celsius degrees; linking decompositional stages or insect development to ADDs allows temperature changes to be taken into account when computing time since death.

acetabulum. The "socket" in the hip, within which the **femoral head** moves.

adipocere. Literally, "grave wax," a greasy or soapy substance formed when fatty tissue decomposes in a moist environment.

antemortem. Before death.

anterior. Toward the front (of the body).

auricular surface. The surface of the hipbone in the area of the **sacroiliac** joint.

autolysis. Literally, "self-digestion," the breakdown of the body's soft tissues as a result of intracellular chemical changes.

autopsy. A **postmortem** examination by a forensic pathologist.

blowfly. Any of several iridescent green or blue flies in the family Calliphoridae that colonize recently deceased bodies, laying eggs in orifices or wounds; the eggs hatch into **maggots** that feed on the soft tissues.

Blumensaat's line. An interior seam in the **femur** just above the knee, named in honor of the German physician who discovered it, now used by anthropologists to help distinguish Negroid femurs from other femurs.

calcaneus. The heel bone, the largest bone of the foot.

cervical. In the region of the neck.

clavicle. Collarbone.

coccyx. The "tailbone," consisting of the lowest (distal) several **vertebrae,** ranging in number from three to five vertebrae.

condyle. A rounded projection or end of a bone, usually where it joins another bone (as, for example, the condyles of the **femur** and of the **tibia** form the "hinge" of the knee).

coronal suture. Joint in the **cranium** running across the top of the head from one side (**parietal**) to the other.

coroner. An official who investigates and certifies deaths; a coroner may or may not have medical training.

cranium. Skull.

cremains. Human remains that have been cremated.

crenulated. Wavy, notched, or scalloped in form; in anthropology, generally used to describe the upper surfaces of the molars of Negroid individuals.

decomposition. The decay or disintegration of the body.

distal. Far; in bones, farther from the center of the body (opposite of **proximal**), as in "the distal end of the femur."

entomologist. A scientist whose specialty is insects.

epiphysis (plural: **epiphyses**). A part of a bone, usually the end, separated from the central portion or shaft by cartilage; specific **epiphyses** ossify at consistent, predictable times, making them important markers of skeletal development or age.

external occipital protuberance. The bony bump at the base of the **occipital** bone of the skull, usually prominent in males but not in females.

femoral head. The "ball" at the **proximal** end of the **femur.**

femur (plural: **femora**). The thighbone.

fibula. The smaller, **lateral** bone of the lower leg.

floater. A body found decomposing in water.

foramen. An opening or hole (in a bone).

foramen magnum. The large opening at the bottom of the **occipital** bone through which the brain stem and spinal cord emerge.

frontal. The bone forming the forehead and upper edges of the eye **orbits.**

greater trochanter. The larger, **lateral epiphysis,** just below the head of the **femur.**

humerus (plural: **humerii**). The bone of the upper arm.

hyoid. Small, **U**-shaped bone at the front of the neck, often broken in cases of strangulation.

ilium. Broad, upper portion of the hipbone, or **innominate.**

innominate. Hipbone, formed by the fusion of the **ilium, ischium,** and **pubic bone.**

instar. Any of the three developmental stages of the **maggot** (first instar, second instar, third instar), distinguished from one another by specific anatomical features, and helpful in pinpointing time since death.

ischium. Lower portion of the hipbone; the part you sit on.

lateral. Toward the side (of the body); the opposite of **medial.**

lesser trochanter. The smaller, **medial epiphysis,** just below the head of the **femur.**

maggot. The caterpillarlike larva of a fly.

mandible. The lower jawbone.

maxilla. The upper jawbone.

medial. Toward the center (of the body); opposite of **lateral.**

medical examiner (ME). A physician who works with law enforcement officers to determine cause of death.

metatarsal. Literally, "beyond instep"; any of the five long bones of the foot
located between the ankle and toes.

ninhydrin. A chemical used to reveal latent human fingerprints; when it reacts with the oils in fingerprints, it turns purple.

occipital. The bone forming the back and base of the skull.

orbit. The bony socket that cradles the eyeball.

ossify. To turn to bone; at birth, the skeleton is formed by cartilage, which
gradually ossifies as calcium and other minerals reinforce it.

osteoarthritic lipping. A degenerative, aging-related process in which joint
surfaces acquire jagged edges through the buildup of additional bony material.

osteology. Literally, "bone science"; the study of bones.

parietal. Literally, "of a wall"; the bone forming either side of the skull.

pathologist. A physician specializing in disease, particularly diseased tissues
and organs; forensic pathologists perform autopsies to determine cause
and manner of death.

pelvis. Literally, "basin"; the pelvis is the structure formed by the **innomi-
nates** and the **sacrum.**

perimortem. At or around the time of death.

phalanges. Bones of the fingers and toes.

postcranial. Below the **cranium,** generally referring to the postcranial skeleton (that is, everything from the neck down).

posterior. Toward the rear (of the body).

postmortem. After death.

proximal. Near; in bones, close to the center of the body (opposite of **distal**),
as in "the proximal end of the femur."

pubic bone (pubis). Anterior portion of **innominate,** where the two hipbones meet at the midline of the abdomen.

pubic symphysis. The junction at the midline of the **pelvis** where the left and right pubic bones meet; the features of the pubic symphysis reveal much about skeletal age.

pupa (plural: **pupae**). Insect in transition from larval stage to adult stage.

puparia. Hard, cocoonlike shells in which insect larvae mature into adults; blowfly puparia are often found by the thousands on or near decomposed bodies or skeletons.

putrefaction. Decomposition of the body's soft tissues, especially by bacteria.

radius. The **lateral** (thumb-side) bone of the forearm.

sacroiliac crest. The seam in the hipbone—normally wide, raised, and prominent in adult females—where the **sacrum** is joined to the **ilium.**

sacrum. Literally, "holy bone"; a triangular bone formed by the fusion of three to five sacral vertebrae; the sacrum is the **posterior** part of the **pelvis.**

scapula. The shoulder blade.

sciatic notch. Gap in the hipbone through which the sciatic nerve passes when it emerges from the lower spine; wider in females than in males.

sphenoid. A U-shaped bone making up the middle floor of the skull.

sternum. The breastbone.

suture. In the context of this book, any of several joints in the skull.

temporal. The bone surrounding the ear.

time since death (TSD). The **postmortem** interval between death and discovery.

thoracic. In the region of the chest.

tibia. The larger, **medial** bone of the lower leg: the "shinbone."

ulna. The **medial** bone of the forearm: the one that includes the sharp bump of the elbow.

vertebra (plural: **vertebrae**). A bone of the spinal column.

zygomatic. The cheekbone.

Acknowledgments

MANY THOUSANDS of people have contributed to making this book a reality. First, I want to thank my mother, the late Jennie Bass, for being a guiding light, especially after the death of my father, all the way up to her death in 1997 at the age of ninety-five. Second, I have been blessed with three wonderful wives (not at the same time, mind you): Ann Owen, who was the mother of our three sons; Annette Blackbourne, who was a wise adviser at work and a great comfort after Ann's death in 1993; and Carol Miles, who has known me since childhood. Carol, who knew both Ann and Annette, came to Knoxville to take care of me when Annette died in 1997; thankfully, she has been here ever since.

I owe a big debt to the thousands—no, tens of thousands—of students who took my classes at the universities of Pennsylvania, Nebraska, Kansas, and Tennessee, and who garnered me for many teaching awards. I've always said that I've had two families: my biological family of three sons and my academic family of all the graduate students who made this pioneering research possible, many of whom you've met in these chapters. I also thank Donna Patton Griffin, one of the able secretaries in the University of Tennessee Anthropology Department, who has typed reports and kept records of hundreds of forensic cases during my years at the university. The Body Farm would not have been a reality without continuous support from the administrators of the University of Tennessee. From the deans in the College of Arts and Sciences, which includes the Anthropology Department, through the chancellors of the Knoxville campus, to the presidents of the statewide UT system, I have received nothing but the greatest support. It's nice to work in an environment where you respect and admire your superiors.

Almost every television crime series portrays friction or conflict between forensic scientists and the police, district attorneys, medical examiners, or coroners they work with. In my fifty years of working with members of local, state, national, and international law enforcement agencies, though, I don't recall a single bad encounter with any of them. I thank them all for the many things they've taught me about arson investigation, ballistics, criminal justice, and other fields I've had to learn in on-the-job training.

I especially want to thank my three sons, Charlie, Billy, and Jim Bass, who have always given me strength, but particularly after the deaths of Ann and Annette. All three of my sons have been very successful: Those education dollars were clearly well spent after all!

Last but not least, I want to thank Jon Jefferson, whose writing has helped make this an engaging story. Jon has become a true friend and a member of the Bass family.

—W.M.B. III

As Goethe once said (a bit more elegantly), the instant you burn your bridges and fling yourself at something, magic happens: Providence moves, doors open, coincidences add up to destiny. This book bears that out. Long before I met her, Cindy Robinson had the foresight to study with a memorable professor; twenty years later she shared her stories of Dr. Bass and his Body Farm with me, a guy who'd had the foresight and good fortune to marry her in the meantime. The best reader and keenest critic I know, Cindy helped me make this book far better.

Many people accompanied and encouraged me on my trek through the realms of the dead, including the two who brought me into the land of the living. Bill and Gloria Jefferson never dreamed this is where their son would end up, but they've remained interested and encouraging throughout my meandering career. So have my children, Ben and Anna, who likewise seem partial to roads less traveled.

My close friend and fellow journalist Steven Keeva published my first story about Dr. Bass and the murder cases he has helped solve. Steve also opened the door that led to my forensic documentaries for the National Geographic Society, and he offered, time and again, faith and hope when my own ran dry. So did John Hoover, a good friend, great listener, and wise counselor. Other mainstays have been my beer buddies from the Wednesday Night Prayer Group: John Craig, J.J. Rochelle, Wendy Smith, and David Brill. David, a fine writer and generous friend, introduced me to Giles Anderson, agent extraordinaire, whose energy and enthusiasm for this project have been inspiring and contagious. Giles, in turn, brought us Danny Baror, our superb international agent.

David Highfill, our editor at Putnam, stepped far out onto a limb to help us bring about the book we so blithely promised him in the first place. Robert Roper has advised me well and guided me past many pitfalls. Nancy Young generously loaned me her cozy cabin and the Carolina mountains when I needed to get away from a myriad of excuses not to write.

Patricia Cornwell's contribution, in spotlighting forensic science in general and the Body Farm in particular, can't be overstated. Her flood tide lifted our boat as surely as her helicopter lifted our spirits that gray day when she hovered with us in the treetops above the Body Farm.

Most of all, I thank Bill Bass and his lovely wife, Carol—a gracious hostess and a lively lunch companion. Bill first suggested this book three years ago; lucky me, to be his collaborator. Working with him has been not just a rare privilege but a constant delight. One of the world's foremost scientists, he's also one of its most humble, honest, and honorable human beings. Unfailingly cheerful, ever enthusiastic, always affirming, Dr. Bass possesses—pacemaker notwithstanding—one of the best hearts to be found anywhere on this beautiful planet brimming with Life.

—J.W.J.

Index